Challenges in Geometry

Challenges in Geometry

for Mathematical Olympians Past and Present

CHRISTOPHER J. BRADLEY

OXFORD
UNIVERSITY PRESS

Great Clarendon Street, Oxford OX2 6DP

Oxford University Press is a department of the University of Oxford.
It furthers the University's objective of excellence in research, scholarship,
and education by publishing worldwide in

Oxford New York

Auckland Cape Town Dar es Salaam Hong Kong Karachi
Kuala Lumpur Madrid Melbourne Mexico City Nairobi
New Delhi Shanghai Taipei Toronto

With offices in

Argentina Austria Brazil Chile Czech Republic France Greece
Guatemala Hungary Italy Japan South Korea Poland Portugal
Singapore Switzerland Thailand Turkey Ukraine Vietnam

Oxford is a registered trade mark of Oxford University Press
in the UK and in certain other countries

Published in the United States
by Oxford University Press Inc., New York

First published 2005

British Library Cataloguing in Publication Data
Data available

Library of Congress Cataloging in Publication Data
Data available

ISBN 0–19–856691–3 9780198566915
ISBN 0–19–856692–1 (pbk) 9780198566922

1 3 5 7 9 10 8 6 4 2

Typeset by Julie M. Harris

Printed in Great Britain
on acid-free paper by
Biddles Ltd., King's Lynn, Norfolk

Preface

This book is written for students who are interested in geometry and number theory, for those involved in Mathematical Olympiads, and for teachers in universities and schools. It is more of a geometry book than a book about integers and contains, among other things, a full account of the properties of triangles and circles normally associated with an advanced course of Euclidean geometry. The restriction to configurations in which various lengths are required to have integer values provides a natural and appealing link between elementary geometry and interesting problems involving Diophantine equations. Though the content is mostly elementary, some of the results would appear to be new. The book is not designed for any particular course of study, but is suitable as additional reading for undergraduates studying these topics, for students preparing for competitions, and for other mathematically advanced high school students.

During the last thirteen years I have been closely involved in the preparation of the United Kingdom team for the International Mathematical Olympiad, of which I was Deputy Leader for three years. The content, therefore, reflects interests developed during these years.

Though few of the problems treated in the book could ever have been set in an Olympiad competition (they are mostly too long and detailed), the techniques involved are precisely those suitable for developing the problem-solving skills needed for competitions. The book also includes a number of topics of a geometrical nature in which integers appear, that are not normally included in a course primarily devoted to Euclidean geometry; for example, there are chapters on polygonal numbers and on methods for obtaining rational and integer points on curves.

I have had two most enjoyable careers. My first was as a lecturer at Oxford University, where my research interests were also in algebra and geometry, and where I had the good fortune to be associated with Professor Charles Coulson and Dr Simon Altmann. The latter was my research supervisor when I was a graduate student and I owe a great deal to his care and enthusiasm. During my years in Oxford I engaged in a major project with my friend Arthur Cracknell, who later became a professor of Physics at the University of Dundee. This project resulted in a book entitled *The mathematical theory of symmetry in solids*, Clarendon Press, Oxford (1972), a classification of the irreducible representations of the 230 space groups.

I left Oxford in 1977 to become a schoolteacher, first at Christ's Hospital, Horsham and later at Clifton College, Bristol. I am grateful to colleagues at both of these schools for their help and encouragement. I retired from full-time work in 1998 and since then there has been time for writing. Since 1990 I have had the privilege to be associated with a number of inspiring colleagues, who have encouraged and challenged me intellectually in ways that I did not anticipate when I became a schoolteacher. These include the various leaders of the UK International Mathematical Olympiad team, Dr Tony Gardiner, Professor Adam McBride, Dr Imre Leader, and Dr Geoff Smith. Perhaps the greatest geometrical influence, however, has been Dr David Monk, who helped to train the UK team for over thirty years, and whose contributions have been immense. Tony Gardiner has spent much time and effort in helping me prepare the manuscript for this book, and I am grateful to him for numerous suggestions for improvements in the style and for the removal of certain ambiguities and conceptual errors. Any remaining errors are entirely my responsibility.

I should also like to thank Dr Kevin Buzzard of the Mathematics Department at Imperial College, London for consultations and assistance with two of the problems in number theory. Thanks are also due to my nephew, Dr Jeremy Bradley, of the Department of Computing at Imperial College, London for help with some of the computational problems in the book and with various pieces of technical help during the course of preparing the manuscript.

I am indebted to readers of the Oxford University Press for invaluable comments and suggestions. I am also extremely grateful to the staff of the Oxford University Press for their unfailing help and encouragement during production, particularly to Kate Pullen, the Assistant Commissioning Editor, whose help and courtesy smoothed my path in the months prior to delivering the final manuscript.

Bristol C. J. B.
July 2004

Contents

Glossary of symbols

The following symbols are repeatedly used in connection with a triangle ABC.

A, B, C	The vertices of the triangle ABC.
$\angle BAC$, $\angle CBA$, $\angle ACB$ or $\angle A$, $\angle B$, $\angle C$	The angles of the triangle ABC.
a, b, c	$a = BC$, $b = CA$, and $c = AB$ are the side lengths of the triangle ABC.
L, M, N	The midpoints of BC, CA, and AB, respectively.
D, E, F	The feet of the altitudes from A, B, and C, respectively.
O	The circumcentre of the triangle ABC.
G	The centroid of the triangle ABC.
H	The orthocentre of the triangle ABC.
I	The incentre of the triangle ABC.
I_1, I_2, I_3	The excentres opposite A, B, and C, respectively.
U, V, W	The points where the internal angle bisectors meet BC, CA, and AB, respectively.
X, Y, Z	The points where the incircle touches BC, CA, and AB, respectively.
T	The nine-point centre.
S	The symmedian point of the triangle ABC.
J	The centre of mass of a uniform wire framework in the shape of the triangle ABC.
R	The radius of the circumcircle.
r	The radius of the incircle.
s	$s = \frac{1}{2}(a + b + c)$ is the semi-perimeter of the triangle ABC.
r_1, r_2, r_3	The radii of the excircles opposite A, B, and C, respectively.
$[ABC]$, $[PQR]$	The areas of the triangles ABC and PQR, respectively.
$[X_1X_2 \cdots X_n]$	The area of the polygon $X_1X_2 \cdots X_n$.

Occasionally, we use some of these symbols to denote other points or quantities, but when this happens it is always made clear in the text. For example, L, M, and N are also used as the points on BC, CA, and AB where the Cevians through a general point P meet BC, CA, and AB, respectively. L, M, and N are also used to denote the feet of the perpendiculars from an arbitrary point P onto the sides BC, CA, and AB, respectively. U, V, and W are also used to denote the midpoints of AH, BH, and CH, respectively, where H is the orthocentre.

1 Integer-sided triangles

To say that a triangle has one side of integer length or that a circle has integer radius is not mathematically significant, as the unit of length can always be adjusted so that this is the case. To say that a triangle has all its sides of integer length is mathematically significant. It means that, whatever the unit of length, the ratio of any pair of side lengths is a rational number. However, even if significant, it is scarcely interesting. This is because, if you are given three positive integers such that the greatest is smaller than the sum of the other two, then you can always construct a triangle having sides with these integer lengths. Integer-sided triangles become mathematically interesting only when some further condition is imposed.

In this first chapter we treat a number of basic problems involving integer-sided triangles, when an additional property is introduced.

A Babylonian tablet confirms that geometers of that era, about 3500 years ago, were aware of the existence of right-angled triangles having integer sides, and may well have had some method for constructing them based on the sexagesimal arithmetic they used. Problems of all sorts involving integers have always been regarded as fascinating and not only by professional mathematicians. Witness the general interest aroused by the solution of Fermat's last theorem.

As seems fitting, since it is such an ancient problem, we start with an account of those integer-sided triangles that have an angle of $90°$. Next we show how to obtain all integer-sided triangles with angles of either $60°$ or $120°$. It would be possible to consider integer-sided triangles in which the angles have cosines that are equal to rational numbers other than 1, $\frac{1}{2}$, or $-\frac{1}{2}$. However, the angles $60°$, $90°$, and $120°$ are special as they feature in the rectangular and hexagonal lattices.

We then consider triangles with integer sides and integer area, and towards the end of the chapter we investigate geometrical figures in which the area and perimeter are related. There is also a section on the rectangular box, which is appropriate to consider at an early stage because of its connection with integer-sided right-angled triangles.

For the most part we deal with configurations in which lengths have integer values, but occasionally we relax this condition and just require them to have rational values. When this is done, it is done simply as a matter of convenience. It is not a restriction because a figure with a finite number of lengths that are rational may always be magnified so that they become integers, the enlargement factor being the least common multiple of all the denominators.

1.1 Integer-sided right-angled triangles

Theorem 1.1.1 (Pythagoras) *ABC is a triangle with $\angle BCA = 90°$ if and only if $a^2 + b^2 = c^2$, where $a = BC$, $b = CA$, and $c = AB$.* □

No proof is needed here.

Theorem 1.1.2 *Suppose that a, b, and c are positive integers with no common factor, that $a^2 + b^2 = c^2$, and that a and b are coprime. Then a and b have opposite parity, so b may be chosen to be even; and with this choice there exist coprime positive integers u and v of opposite parity with $u > v$ such that*

$$a = u^2 - v^2\,, \quad b = 2uv\,, \quad c = u^2 + v^2\,. \tag{1.1.1}$$

Proof The integers a and b cannot both be even, for then a, b, and c would have a common factor of 2. The integers a and b cannot both be odd, since all odd squares are equal to 1 (mod 4) and $c^2 = 2$ (mod 4) is impossible. Suppose then that a is odd and b is even. Then c is also odd. We have $b^2 = c^2 - a^2 = (c-a)(c+a)$. Now, since a and c are both odd, $c - a$ and $c + a$ have a factor of 2 in common and cannot both be divisible by 4. But a and c themselves have no factor in common, so $c - a$ and $c + a$ have no factor other than 2 in common. Hence each must be twice a perfect square, the squares having no factor in common. Writing $c + a = 2u^2$ and $c - a = 2v^2$, we find $b^2 = 4u^2v^2$ and $b = 2uv$, $c = u^2 + v^2$, and $a = u^2 - v^2$, where, since a and c are odd and coprime, u and v must be of opposite parity and coprime. Also, u and v may be chosen to be both positive, since b is positive and $u > v$, since a is positive.

Note that $(u^2 - v^2)^2 + (2uv)^2 = (u^2 + v^2)^2$, so that condition (1.1.1) is sufficient as well as necessary for such an integer triple (a, b, c) to exist. □

These integer triples (a, b, c) are called *primitive Pythagorean triples*. They are *Pythagorean* because a, b, and c may then be chosen to be the integer sides of a right-angled triangle and *primitive* because a, b, and c have no common factor. The general solution of $a^2 + b^2 = c^2$ in integers is then found by enlargement by a scale factor k:

$$a = k(u^2 - v^2)\,, \quad b = 2kuv\,, \quad c = k(u^2 + v^2)\,, \tag{1.1.2}$$

where k is a positive integer. The simplest example with $k = 1$, $u = 2$, and $v = 1$ is illustrated in Fig. 1.1.

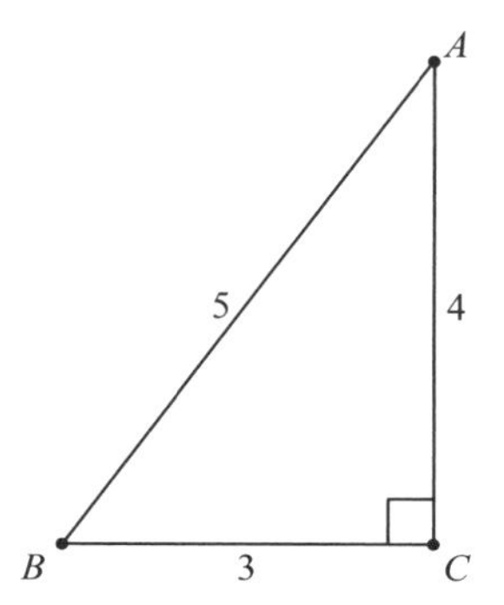

Fig. 1.1 Pythagoras' theorem.

Exercises 1.1

1.1.1 Generalise the patterns appearing in Table 1.1 when (a) $v = 1$, and (b) $u = v + 1$.

1.1.2 Prove that one member of a Pythagorean triple is always divisible by 5 and that the area of an integer-sided right-angled triangle is always divisible by 6.

1.1.3 Find all integer-sided right-angled triangles with hypotenuse 145.

1.1.4 Prove that if c is the hypotenuse of a primitive integer-sided right-angled triangle then c^2 is also.

1.1.5 Let (a, b, c) be a primitive Pythagorean triple. Prove that there exists an infinite number of sets of integers l, m, and n such that

$$a = -lb + mc\,, \quad b = la - nc\,, \quad c = ma - nb \tag{1.1.3}$$

and

$$m^2 + n^2 = 1 + l^2\,. \tag{1.1.4}$$

Conversely, prove that if l, m, and n are integers satisfying eqn (1.1.4) then there exists a primitive Pythagorean triple (a, b, c) satisfying eqn (1.1.3).

1.1.6 Prove that there are an infinite number of primitive Pythagorean triples in which $|a - b| = 1$. Explain how this result is related to finding rational approximations of $\sqrt{2}$.

1.1.7 Is it possible to find an infinite set of points in the plane, not all on the same straight line, such that the distance between every pair of points is rational?

Table 1.1 Primitive Pythagorean triples with $c < 100$.

u	v	$a = u^2 - v^2$	$b = 2uv$	$c = u^2 + v^2$
2	1	3	4	5
3	2	5	12	13
4	1	15	8	17
4	3	7	24	25
5	2	21	20	29
5	4	9	40	41
6	1	35	12	37
6	5	11	60	61
7	2	45	28	53
7	4	33	56	65
7	6	13	84	85
8	1	63	16	65
8	3	55	48	73
8	5	39	80	89
(8	7	15	112	113)
9	2	77	36	85
9	4	65	72	97

1.2 Integer-sided triangles with angles of $60°$ and $120°$

Suppose now that a, b, and c are the side lengths of an integer-sided triangle with angles of either $60°$ or $120°$. Note that $\cos 60° = \frac{1}{2}$ and $\cos 120° = -\frac{1}{2}$. It follows from the cosine rule that if angle $C = 60°$ then $c^2 = a^2 - ab + b^2$, and if angle $C = 120°$ then $c^2 = a^2 + ab + b^2$.

We first show how to obtain all solutions in nonzero integers a, b, and c of the equation

$$c^2 = a^2 - ab + b^2\,, \tag{1.2.1}$$

without regard to the geometrical application. In the analysis that follows it turns out that we can always ensure that c is positive. Then a solution of eqn (1.2.1) in which precisely one of a or b turns out to be negative is a solution with positive a and b of the equation

$$c^2 = a^2 + ab + b^2 \tag{1.2.2}$$

by a change of sign of a or b, as appropriate. If both of a and b turn out to be negative, then a change of sign of both of them gives a solution of eqn (1.2.1) in which all of a, b, and c are positive. In this way, the positive solutions of both equations may be obtained simultaneously.

The method is to find a non-singular unimodular linear transformation from a, b, and c to new integer variables u, v, and w to provide an equation that is linear in each of the variables u, v, and w. The transformed equation can then be solved to find w in terms of u and v. Finally, by eliminating w, the variables a, b, and c can be expressed in terms of the two-parameter system u and v. The method works with most of the homogeneous quadratic Diophantine equations that I have encountered and it is one that I use several times in this text. There are many non-singular linear transformations that will do the job.

Here a suitable transformation is $a = u - w$, $b = v - w$, and $c = u + v - w$, and then substitution into eqn (1.2.1) gives $3uv = w(u+v)$. Hence $w = 3uv/(u+v)$. Now, substituting back for w and multiplying up by $u+v$, we obtain the two-parameter solution

$$\begin{aligned} a &= u^2 - 2uv\,, \\ b &= v^2 - 2uv\,, \\ c &= u^2 - uv + v^2\,. \end{aligned} \tag{1.2.3}$$

Certain points need to be made about the method. Firstly, it does not always lead to a solution in which a, b, and c are coprime. For example, $u = 5$ and $v = 1$ gives $a = 15$, $b = -9$, and $c = 21$. This provides the primitive solution $a = 5$, $b = 3$, and $c = 7$ to eqn (1.2.2). Secondly, solutions get repeated. For example, $u = 3$ and $v = 1$ leads to the solution $a = 3$, $b = 5$, and $c = 7$ of eqn (1.2.2). To get all solutions one has to multiply each of the expressions above for a, b, and c by any positive integer k. The cases $u = 2v$ and $v = 2u$ give only trivial solutions and must be excluded. The case $u = 1$ and $v = 1$ gives $a = -1$, $b = -1$, and $c = 1$, which on changing the signs of both a and b gives the equilateral triangle solution $a = b = c = 1$ of eqn (1.2.1). The case $u = 1$ and $v = -1$ gives the solution $a = b = c = 3$ of eqn (1.2.1).

The important question is whether all solutions arise using the method. The answer is 'yes'. To see this, note that the inverse transformation is $u = c - b$, $v = c - a$, and $w = c - a - b$, so that for any a, b, and c ($c > 0$) satisfying eqn (1.2.1) values of u, v, and w always exist. However, since we multiply up by $u + v$ the solution for a, b, and c may be a multiple of the specified solution. For example, consider the solution $a = 8$, $b = 15$, and $c = 13$ of eqn (1.2.1). From the inverse transformation we have $u = -2$, $v = 5$, and $w = -10$. Substituting back into our expressions for a, b, and c in terms of u and v we get $a = 24$, $b = 45$, and $c = 39$. It is easy to see from the transformation that a, b, and c are coprime provided that u and v are coprime and that 3 is not a factor of $u + v$. When 3 is a factor of $u + v$ all that happens is that a, b, and c have a common factor of 3, so the difficulty is a minor one. In Table 1.2 we give the first few solutions to eqns (1.2.1) and (1.2.2) for triples $\{a, b, c\}$ for triangles with $C = 60^\circ$ and $[a, b, c]$ for triangles with $C = 120^\circ$. See Figs 1.2 and 1.3 for examples of each of these cases.

Table 1.2 The following triangles have $C = 60°$ or $120°$ and corresponding triples are labelled $\{a, b, c\}$ or $[a, b, c]$, respectively.

$\{1, 1, 1\}$;
$\{8, 5, 7\}$, $\{8, 3, 7\}$, $\{5, 8, 7\}$, $[3, 5, 7]$;
$\{15, 7, 13\}$, $\{15, 8, 13\}$, $\{7, 15, 13\}$, $[8, 7, 13]$;
$\{24, 9, 21\}$, $\{24, 15, 21\}$, $\{9, 24, 21\}$, $[15, 9, 21]$;
$\{21, 16, 19\}$, $\{21, 5, 19\}$, $\{16, 21, 19\}$, $[5, 16, 19]$;
$\{35, 11, 31\}$, $\{35, 24, 31\}$, $\{11, 35, 31\}$, $[24, 11, 31]$;
$\{48, 13, 43\}$, $\{48, 35, 43\}$, $\{13, 48, 43\}$, $[35, 13, 43]$;
$\{45, 24, 39\}$, $\{45, 21, 39\}$, $\{24, 45, 39\}$, $[21, 24, 39]$;
$\{40, 33, 37\}$, $\{40, 7, 37\}$, $\{33, 40, 37\}$, $[7, 33, 37]$

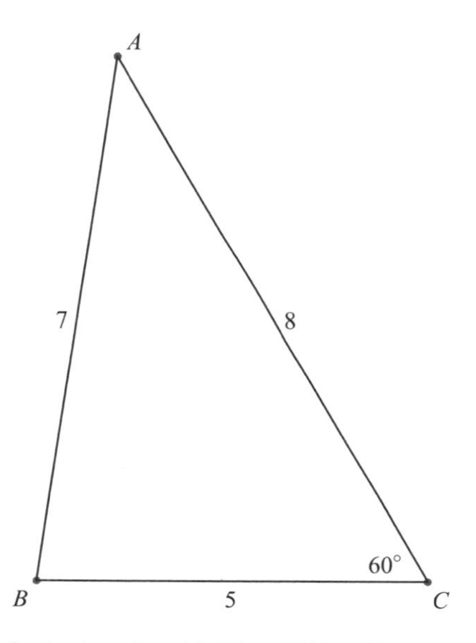

Fig. 1.2 A triangle with $C = 60°$ and integer sides.

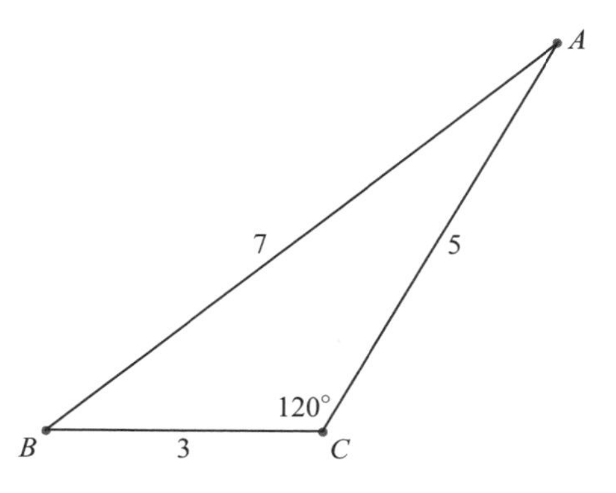

Fig. 1.3 A triangle with $C = 120°$ and integer sides.

Exercises 1.2

1.2.1 Explain algebraically the repeats in the list of triangles with angle $C = 60°$ in Table 1.2.

1.2.2 Explain geometrically why, if $[a, b, c]$ is a triple with angle $C = 120°$, then $\{a + b, b, c\}$ is a triple with angle $C = 60°$.

1.2.3 Find all integer-sided triangles with angle $C = 120°$ and $c = 91$.

1.2.4 Prove the statement in the text, that a, b, and c given by eqn (1.2.3) are coprime provided that u and v are coprime and that 3 is not a factor of $u + v$.

1.3 Heron triangles

A Heron triangle is often defined to be one with rational side lengths and rational area. Clearly, a triangle similar to a Heron triangle is also a Heron triangle, provided that the scale factor is rational, and indeed any Heron triangle is similar to a Heron triangle with integer side lengths and area. We shall adopt this more restrictive definition and insist that *all of our Heron triangles have integer side lengths and integer area*. Indeed, since the altitudes of a Heron triangle must have rational length, we can, if needed, find a Heron triangle similar to a given one, for which one (or even all) of the altitudes has integer length. In what follows we insist that *the altitude from A is of integer length*.

It is our purpose, in this section, to give a parametric classification of Heron triangles as defined above, by means of the theory of Section 1.1. In Section 7.5 we describe an alternative and more subtle approach to the classification of Heron triangles.

An integer-sided right-angled triangle is trivially a Heron triangle and these have already been dealt with in Section 1.1.

From the cosine rule, $\cos C = (a^2 + b^2 - c^2)/2ab$, etc., it follows that for an integer-sided triangle the cosine of each angle is rational. From this observation alone, it is evident that each acute-angled Heron triangle ABC is the union of two integer-sided right-angled triangles ABD and ACD or, if angle B (or C) is obtuse, the difference of two integer-sided right-angled triangles ACD and ABD. It would be possible to dispense with Heron triangles that are the difference of two right-angled triangles if we were to insist that any obtuse angle is placed at the vertex A, but we choose not to do this.

The following formulae hold for the area $[ABC]$ of triangle ABC:

$$\begin{aligned}[ABC] &= \frac{1}{2}bc\sin A = \frac{1}{2}ca\sin B = \frac{1}{2}ab\sin C \\ &= \frac{abc}{4R} = \{s(s-a)(s-b)(s-c)\}^{1/2} \qquad (1.3.1) \\ &= \frac{1}{2}ad = \frac{1}{2}be = \frac{1}{2}cf = rs\,.\end{aligned}$$

Here R is the circumradius, r is the inradius, $s = \frac{1}{2}(a + b + c)$ is the semi-perimeter, and d, e, and f are the lengths of the altitudes AD, BE, and CF, respectively. It follows, in particular, that for a Heron triangle the sine of each angle is rational.

As a first example, take $AB = 13$, $AC = 15$, $BC = 14$, $AD = 12$, $BD = 5$, $CD = 9$, and $[ABC] = 84$. This is the union of a $(5, 12, 13)$ triangle and a threefold enlargement of a $(3, 4, 5)$ triangle. The triangle with $BC = 4$, $AB = 13$, $AC = 15$, and $[ABC] = 24$ is the difference of these triangles. These examples, illustrated in Fig. 1.4, involve a component right-angled triangle that is an enlargement of a primitive right-angled triangle. However, Heron triangles exist that are the union or difference of two primitive Pythagorean triangles, and we now obtain formulae for their side lengths.

Theorem 1.3.1 (Heron triangles with even height and primitive components) *Let ABC be a Heron triangle, as defined above, with integer sides, integer area, and integer altitude AD. If AD is even and the Heron triangle is built from two primitive right-angled triangles, then positive integer parameters w, x, y, and z exist with $wx > yz$ and $wy > xz$ such that $AB = w^2x^2 + y^2z^2$, $AC = w^2y^2 + x^2z^2$, and* **either** *$BC = (w^2 - z^2)(x^2 + y^2)$ and $[ABC] = wxyz(x^2 + y^2)(w^2 - z^2)$ when ABC is acute* **or** *$BC = (w^2 + z^2)|x^2 - y^2|$ and $[ABC] = wxyz(w^2 + z^2)|x^2 - y^2|$ when ABC is obtuse.*

Proof Take the altitude $AD = 2uv = 2pq$; then we have integers w, x, y, and z such that $u = wx$, $v = yz$, $p = wy$, and $q = xz$, where $u > v$ and $p > q$, u and v are

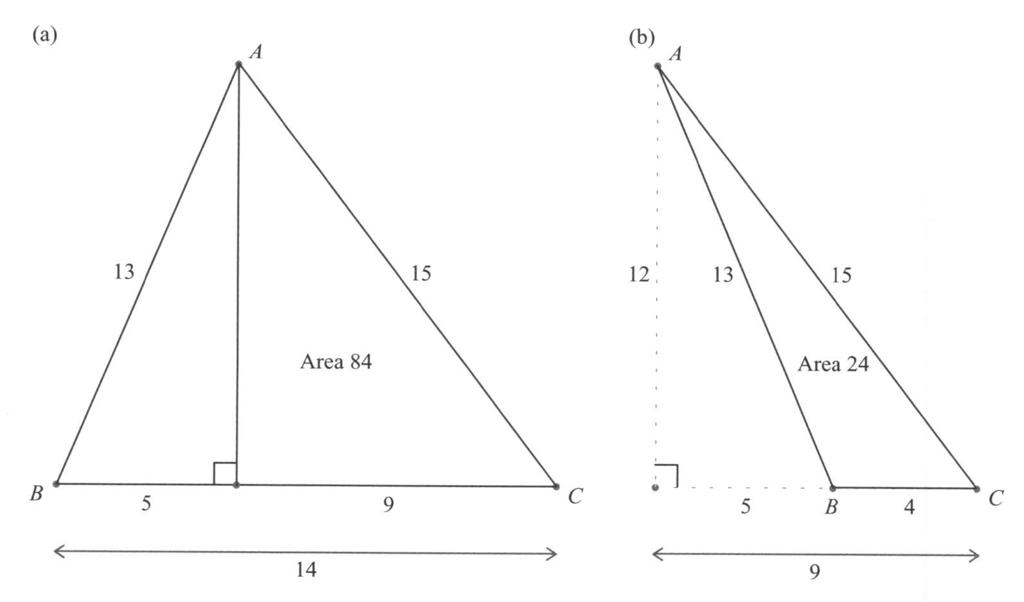

Fig. 1.4 (a) The acute-angled Heron triangle, and (b) the obtuse-angled Heron triangle, with shortest side lengths.

coprime and of opposite parity, and p and q are coprime and of opposite parity. Then $AB = u^2 + v^2$, $AC = p^2 + q^2$, $BD = u^2 - v^2$, $CD = p^2 - q^2$, and $BC = BD + CD$ in the acute case, and $BC = |BD - CD|$ in the obtuse case. □

As an example, take $w = 2$, $x = 3$, $y = 5$, and $z = 1$. Then $u = 6$, $v = 5$, $p = 10$, and $q = 3$. So $AB = 61$, $AC = 109$, and $BC = 11 + 91 = 102$ in the acute case and $BC = 91 - 11 = 80$ in the obtuse case. In both cases $AD = 60$. The acute case is illustrated in Fig. 1.5.

Theorem 1.3.2 (Heron triangles with odd height and primitive components) *With the same hypotheses as Theorem 1.3.1, but with the altitude AD being an odd integer, then positive integer parameters w, x, y, and z exist with $wx > yz$ and $wy > xz$ such that $AB = \frac{1}{2}(w^2x^2 + y^2z^2)$, $AC = \frac{1}{2}(w^2y^2 + x^2z^2)$, and* **either** *$BC = \frac{1}{2}(w^2 - z^2)(x^2 + y^2)$ and $[ABC] = \frac{1}{4}(w^2 - z^2)(x^2 + y^2)wxyz$ when ABC is acute,* **or** *$BC = \frac{1}{2}(w^2 + z^2)|x^2 - y^2|$ and $[ABC] = \frac{1}{4}(w^2 + z^2)|x^2 - y^2|wxyz$ when ABC is obtuse.*

Proof Taking the altitude $AD = u^2 - v^2 = p^2 - q^2$ with $u > v$, $p > q$, u and v coprime and of opposite parity, and p and q coprime and of opposite parity, then we have odd integers w, x, y, and z such that $u = \frac{1}{2}(wx + yz)$, $v = \frac{1}{2}(wx - yz)$, $p = \frac{1}{2}(wy + xz)$, and $q = \frac{1}{2}(wy - xz)$. Then $AB = u^2 + v^2$, $AC = p^2 + q^2$, $BD = 2uv$, $CD = 2pq$, and $BC = BD + CD$ in the acute case and $BC = |BD - CD|$ in the obtuse case. □

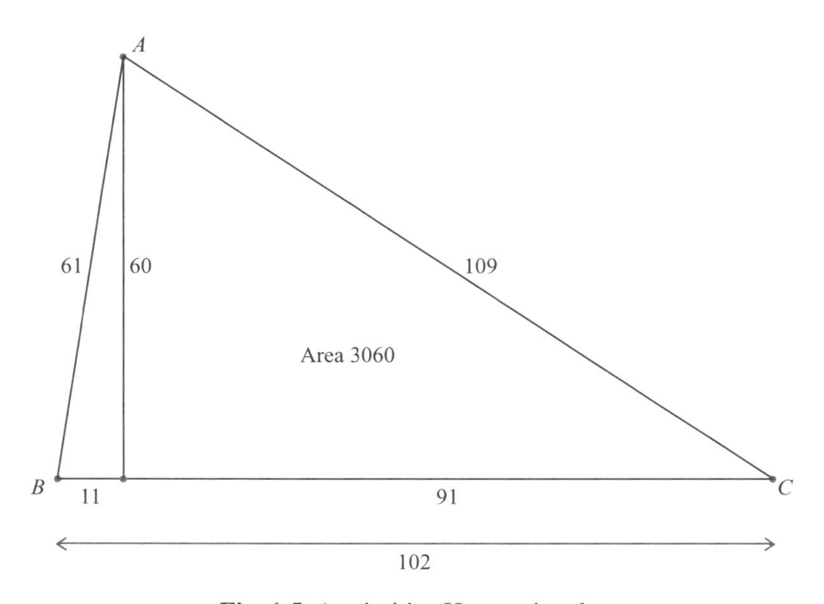

Fig. 1.5 A primitive Heron triangle.

As an example, take $w = 3$, $x = 5$, $y = 7$, and $z = 1$. Then $u = 11$, $v = 4$, $p = 13$, and $q = 8$. So $AB = 137$, $AC = 233$, and $BC = 88 + 208 = 296$ in the acute case and $BC = 208 - 88 = 120$ in the obtuse case. In both cases $AD = 105$.

Heron triangles with component right-angled triangles that are not primitive are even more common. They can be constructed in a similar manner by choosing integers h, k, u, v, p, and q such that **either** $2huv = 2kpq$ or $k(p^2 - q^2)$ **or** $h(u^2 - v^2) = 2kpq$ or $k(p^2 - q^2)$. We give an example of a Heron triangle with non-primitive components in which $h = 3$, $u = 4$, $v = 1$, $k = 2$, $p = 3$, and $q = 2$, so that $2huv = 2kpq$. The values of the parameters h and k show that one component triangle is an enlargement by a factor of 3 of an $(8, 15, 17)$ triangle, and that the other component triangle is an enlargement by a factor of 2 of a $(12, 5, 13)$ triangle. Hence $AB = 51$, $AC = 26$, $AD = 24$, and $BC = 55$ when ABC is acute and $BC = 35$ when ABC is obtuse.

An alternative and instructive method of characterising Heron triangles is as follows. Suppose that $AD = h$, $AB = c$, $AC = b$, $BD = n$, and $CD = m$, where h, b, c, m, and n are integers. We have $h^2 = c^2 - n^2 = b^2 - m^2$, so $c^2 + m^2 = b^2 + n^2$. It follows that integers p, q, r, and s exist so that

$$\begin{aligned} b &= \frac{1}{2}(pr + qs)\,, \\ c &= \frac{1}{2}(pq + rs)\,, \\ m &= \frac{1}{2}(pr - qs)\,, \\ n &= \frac{1}{2}(pq - rs)\,. \end{aligned} \tag{1.3.2}$$

Here, either p, q, r, and s are all odd, or p and s are both even, or q and r are both even (or three of p, q, r, and s are even). Then $h^2 = c^2 - n^2 = pqrs$. Now it is very easy to choose four integers so that their product is a perfect square, and every such choice leads to two Heron triangles, an acute one with $BC = m + n$ or an obtuse one with $BC = |m - n|$. As an example, let $p = 6$, $q = 3$, $r = 4$, and $s = 2$. Then $AB = 13$, $AC = 15$, $AD = 12$, and $BC = 14$ or 4. More generally, one can choose $p = klmw^2$, $q = ktux^2$, $r = ltvy^2$, and $s = muvz^2$.

Note from formulae (1.3.1) that $[ABC] = abc/4R = rs$, where R and r are the circumradius and inradius of the triangle ABC, respectively, and it follows that in a Heron triangle both R and r are rational.

Exercises 1.3

1.3.1 Show that if $\cos B = n/c$ and $\cos C = m/b$, where b, c, m, and n are defined in terms of p, q, r, and s as in the penultimate paragraph of Section 1.3, then

$$\cos A = \frac{ps(q+r)^2 - qr(p-s)^2}{(pq+rs)(pr+qs)}.$$

1.3.2 Prove that there are an infinite number of Heron triangles whose side lengths are consecutive integers.

1.3.3 Find all Heron triangles with an altitude of 40.

1.3.4 Show that an infinite number of Heron triangles exist in which two side lengths are perfect squares.

1.3.5 Prove that, in a Heron triangle in which the sides have no common factor, two of the sides are odd and one is even.

1.4 The rectangular box

We consider three interesting possibilities that arise in connection with a rectangular parallelepiped with integer side lengths a, b, and c.

The first problem, and one to which a complete answer can be given, is whether a, b, and c can be chosen so that the main diagonals are integers. That is, for which positive integers a, b, and c does an integer d exist such that $a^2+b^2+c^2 = d^2$? This is, of course, nothing more than the three-dimensional generalisation of the right-angled triangle problem, and its solution leads to *Pythagorean quartets*, such as $(1, 2, 2, 3)$ and $(2, 3, 6, 7)$. See below for a complete solution to this problem.

The second problem is a more sophisticated one and asks whether a, b, and c can be chosen so that all three face diagonals are integers. That is, for which integers a, b, and c do three integers A, B, and C exist so that $b^2 + c^2 = A^2$, $c^2 + a^2 = B^2$, and $a^2 + b^2 = C^2$? One might imagine that cases would be rare or even non-existent, but the surprising result is that a two-parameter system of solutions exists, which is in 1–1 correspondence with the primitive Pythagorean triples.

Evidence from similar problems (see Sections 2.4 and 6.4) indicates that other parametric systems of solutions probably exist and that sporadic solutions may also exist. The triple with least positive a, b, and c is $(44, 117, 240)$. See below for further details about these results.

The third problem is whether a, b, and c exist so that simultaneously all face diagonals and leading diagonals are integers. This is an open question and it is a very difficult problem, not only because of our incomplete knowledge of the solutions of the second problem, but also because the existence of a fourth equation raises the problem to a higher level of difficulty.

Integer-sided main diagonal

Theorem 1.4.1 *The general solution in positive integers of the equation* $a^2+b^2+c^2=d^2$ *is*

$$\begin{aligned} a &= k(p^2+q^2-u^2-v^2)\,,\\ b &= 2k(pu-qv)\,,\\ c &= 2k(qu+pv)\,,\\ d &= k(p^2+q^2+u^2+v^2)\,, \end{aligned} \tag{1.4.1}$$

where p, q, u, *and* v *have no common factor, one or three of* p, q, u, *and* v *are odd,* $p^2+q^2>u^2+v^2$, $pu>qv$, $qu>-pv$, *and* k *is a positive integer.*

Outline proof As the equation is homogeneous the factor k accounts for any common factor, so we need only consider the case in which a, b, c, and d have no common factor. This means that a, b, and c cannot all be even. Neither can two or three of them be odd, since d^2 cannot equal 2 or 3 (mod 4). Hence we may suppose that a is odd, b and c are even, and d is odd.

Writing the equation as $(b+\mathrm{i}c)(b-\mathrm{i}c)=(d-a)(d+a)$ and working over the Gaussian integers, we may factorise into Gaussian primes and use the property of unique factorisation to obtain $b+\mathrm{i}c=2wz$, $b-\mathrm{i}c=2w^*z^*$, $d+a=2zz^*$, and $d-a=2ww^*$. Putting $z=p+\mathrm{i}q$ and $w=u+\mathrm{i}v$ we now obtain the solution given. In order that a and d should be odd when $k=1$, we must choose precisely one or three of p, q, u, and v to be odd. The other conditions ensure that a and b are positive. For those unfamiliar with the Gaussian integers I recommend Silverman (1997). In chapters 33 and 34 he discusses the units and primes of the Gaussian integers and the property of unique factorisation necessary for the above argument to be made fully complete. □

When a, b, c, and d have no common factor we call the Pythagorean quartet *primitive*. We note that, as every positive integer is the sum of four squares, there is a primitive quartet for almost all odd values of d. Any exception results solely from the fact that at least three of p, q, u, and v must be nonzero and one or three of them odd.

Sometimes a selection leads to a quartet that is not primitive. For example, $p=3$, $q=1$, $u=2$, and $v=1$ leads to $a=5$, $b=10$, $c=10$, and $d=15$. (Cases such as this may arise when p^2+q^2 and u^2+v^2 have a common factor, and it is more trouble than it is worth to exclude these cases.) In Table 1.3 we list the primitive Pythagorean quartets with $d<20$ and in Fig. 1.6 we illustrate one case.

Integer-sided face diagonals

A two-parameter set of solutions is given below in eqn (1.4.2), in which a, b, c, A, B, and C are integers satisfying the equations $b^2+c^2=A^2$, $c^2+a^2=B^2$, and $a^2+b^2=C^2$. A derivation is not possible since other solutions exist. It is left to the reader to verify that the above three equations are satisfied.

Table 1.3 Primitive Pythagorean quartets with $d < 20$.

a	b	c	d
1	2	2	3
3	2	6	7
7	4	4	9
1	8	4	9
9	6	2	11
7	6	6	11
3	4	12	13
11	2	10	15
5	2	14	15
9	12	8	17
1	12	12	17
15	6	10	19
17	6	6	19
1	18	6	19

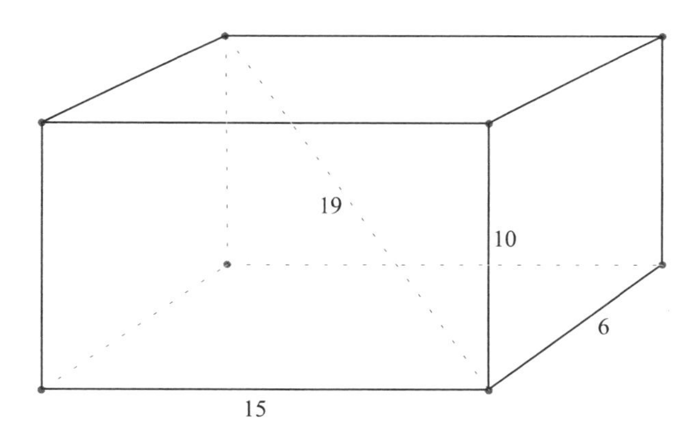

Fig. 1.6 Integer sides and integer diagonal.

The result, apart from a possible common multiplier, is given in terms of any primitive Pythagorean triple (x, y, z) and is as follows:

$$\begin{aligned} a &= y|3x^2 - y^2|, \quad & b &= x|3y^2 - x^2|, \quad & c &= 4xyz, \\ A &= x(5y^2 + x^2), \quad & B &= y(5x^2 + y^2), \quad & C &= z^3. \end{aligned} \tag{1.4.2}$$

As noted earlier, these solutions are in 1–1 correspondence with the primitive Pythagorean triples, which is an intriguing result. The triple $(3, 4, 5)$ corresponds to $a = 44$, $b = 117$, $c = 240$, $A = 267$, $B = 244$, and $C = 125$, see Fig. 1.7. An example of a solution that does not belong to the above family of solutions is $a = 1008$, $b = 1100$, $c = 1155$, $A = 1595$, $B = 1533$, and $C = 1492$. I do not know whether this is part of a system of solutions or whether it is a sporadic solution.

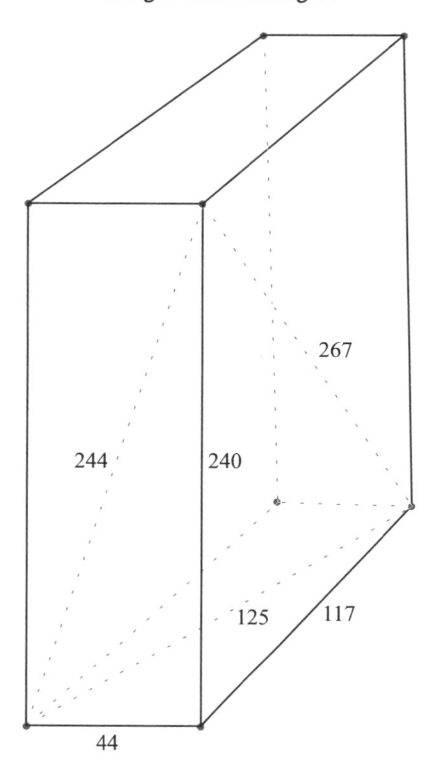

Fig. 1.7 Integer sides and integer face diagonals.

Exercises 1.4

1.4.1 Find all primitive Pythagorean quartets with $d = 25$.

1.4.2 Show that the equation

$$(u^4 - v^4)\sin 2x + 2uv(u^2 + v^2)\cos 2x = 4uv(u^2 - v^2)$$

has one solution given by $\tan x = v/u$. What is the other solution? If u and v are the parameters for a Pythagorean triple as defined in Section 1.1, show that the second value of $\tan x$ gives the parameters for the Pythagorean triple (a, b, C) as defined in the last paragraph of Section 1.4.

1.4.3 Show that primitive Pythagorean quartets exist with $b = a+1$ and $d = c+1$, and that quartets also exist with $b = a - 1$ and $d = c + 1$.

1.4.4 Show that if r, s, t, and u are integers such that $r^2 = tu + us + st$ then (a, b, c, d) is a Pythagorean quartet with $a = r + s$, $b = r + t$, $c = r + u$, and $d = r + s + t + u$. (This result enables one to obtain Pythagorean quartets very easily, as solutions of the equation $r^2 = tu \pmod{s}$ are very easy to construct. See also Section 7.5, where a parametric solution of this equation is given.)

1.5 Integer-related triangles

An *integer-related triangle* is defined to be an integer-sided triangle in which the ratio $[ABC]/(a+b+c)$ is an integer. A triangle is said to be *equable* if this ratio is equal to 1. The problem of finding all equable triangles is a nice problem in integer geometry. However, I disclaim responsibility for the use of the word 'equable', which I do not like. Nor did I like a GCSE coursework task (for UK pupils, aged 15–16) which asked pupils to study 'equable shapes'. My reason for disliking the task was that, because the ratio has the dimension of length, every shape is similar to an equable shape! As far as I am aware, the problem about equable triangles first appeared in a USSR Olympiad, see Shklarsky *et al.* (1993).

Equable triangles

Using Heron's formula for the area, we find that the condition for a triangle to be equable is

$$\{(a+b+c)(b+c-a)(c+a-b)(a+b-c)\}^{1/2} = 4(a+b+c)\,, \qquad (1.5.1)$$

where a, b, and c are positive integers satisfying $b+c > a$, $c+a > b$, and $a+b > c$. Put $b+c-a = l$, $c+a-b = m$, and $a+b-c = n$, so that $a = \frac{1}{2}(m+n)$, $b = \frac{1}{2}(n+l)$, $c = \frac{1}{2}(l+m)$, and $a+b+c = l+m+n$. Note that the triangle inequalities for a, b, and c are satisfied if and only if l, m, and n are positive. In terms of l, m, and n, eqn (1.5.1) reduces to

$$lmn = 16(l+m+n)\,. \qquad (1.5.2)$$

Note that for a, b, and c to be integers l, m, and n have to be positive integers of the same parity, and from eqn (1.5.2) it is evident that they must all be even. So, putting $l = 2p$, $m = 2q$, and $n = 2r$, we have $a = q+r$, $b = r+p$, and $c = p+q$, where p, q, and r satisfy $pqr = 4(p+q+r)$ and p, q, and r are positive integers. Solving for p we obtain

$$p = \frac{4(q+r)}{qr-4}\,. \qquad (1.5.3)$$

Suppose, without loss of generality, that p, q, r is a descending sequence, with possible equalities; then p is not less than $\frac{1}{2}(q+r)$. It now follows from eqn (1.5.3) that $4 < qr$ and $qr \leqslant 12$. It is possible, therefore, to test all possible values of q and r to see from eqn (1.5.3) which values make p integral and at least as great as q. The results and the corresponding sides of the triangle are given in Table 1.4. Up to congruence, there are only five equable triangles, two of them being right-angled and the other three obtuse. These five triangles are shown in Fig. 1.8.

Table 1.4 The five equable triangles.

p	q	r	a	b	c
10	3	2	5	12	13
6	4	2	6	8	10
24	5	1	6	25	29
14	6	1	7	15	20
9	8	1	9	10	17

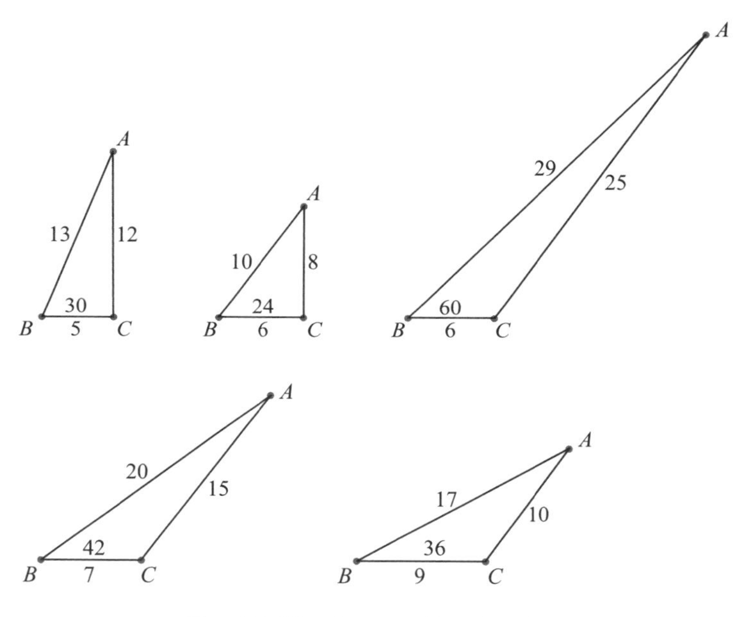

Fig. 1.8 The five equable triangles.

Exercise 1.5

1.5.1 Show that if the triangle ABC is integer-related and $[ABC]/(a+b+c) = 2$ then integers p, q, and r exist such that $p = 16(q+r)/(qr-16)$, where $a = q+r$, $b = r+p$, and $c = p+q$. Hence determine the sides of the eighteen triangles that are integer-related with ratio 2.

1.6 Other integer-related figures

Equable parallelograms

I define an *equable parallelogram* to be a parallelogram with integer sides and integer area, in which the distance between one pair of parallel sides is an integer and the ratio of the area to the perimeter is equal to 1.

Suppose that $ABCD$ is a parallelogram with $AB = CD = b$ and $BC = DA = a$. Suppose further that the distance between the sides AB and CD is equal to h, where a, b, and h are integers and $h < a$. The condition of equal magnitude for area and perimeter is $2(a + b) = hb$, which implies that $(hb - 2a)(h - 2) = 4a$. Writing $h = 2k$, where k is an integer or half an odd integer, gives $(kb - a)(k - 1) = a$, so we may take $kb - a = p$ and $k - 1 = q$, and then $(1 + q)b = p + pq$. In terms of p and k we therefore have $a = p(k - 1)$, $h = 2k$, and $b = p$. If k is an integer then p is an integer. If k is half an odd integer then p must be even. Also, we must have $h < a$, which implies that $2k < p(k - 1)$ or $p > 2k/(k - 1)$. As an example, we may take $k = 5/2$, $p = 4$, $h = 5$, $b = 4$, and $a = 6$, and the acute angle of the parallelogram is equal to $\arcsin(5/6)$, see Fig. 1.9.

Equable rectangular boxes

Equable rectangular boxes are defined to be such that their surface area is equal in magnitude to their volume.

If the dimensions of the box are x, y, and z then the condition for an equable rectangular box is

$$xyz = 2(yz + zx + xy)\,. \tag{1.6.1}$$

This may be rewritten as

$$1 = \frac{2}{x} + \frac{2}{y} + \frac{2}{z}\,. \tag{1.6.2}$$

Suppose that x is the least and z is the greatest dimension (equality being allowed); then eqn (1.6.2) implies that x cannot be greater than 6. Nor can $x = 1$ or 2. So we can solve the problem by trying $x = 3$, 4, 5, and 6 in turn.

When $x = 3$ we have $yz = 6(y + z)$ or $(y - 6)(z - 6) = 36 = 1 \times 36$ or 2×18 or 3×12 or 4×9 or 6×6. This leads to the five solutions $(3, 7, 42)$, $(3, 8, 24)$, $(3, 9, 18)$, $(3, 10, 15)$, and $(3, 12, 12)$. When $x = 4$ we find that $(y - 4)(z - 4) = 16$, leading to three more solutions $(4, 5, 20)$, $(4, 6, 12)$, and $(4, 8, 8)$. The case $x = 5$ gives $(3x - 10)(3y - 10) = 100 = 2 \times 50$ or 5×20. Only the second of these leads to a new solution, namely $(5, 5, 10)$. The case $x = 6$ leads to $(y - 3)(z - 3) = 9$ and

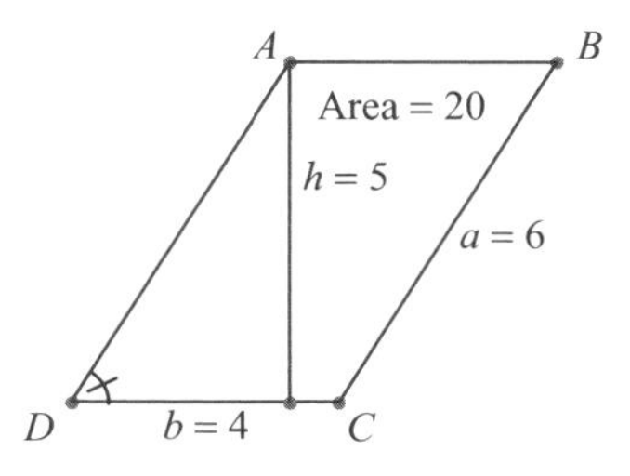

Fig. 1.9 An equable parallelogram with $\angle ADC = \arcsin(5/6)$.

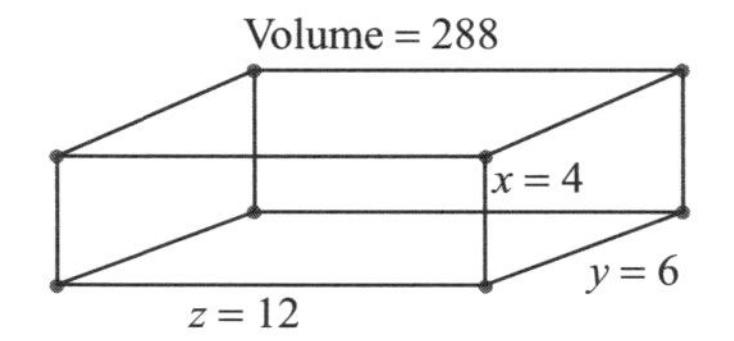

Fig. 1.10 An equable rectangular box.

this provides only one new solution $(6, 6, 6)$. There are ten solutions altogether, one of which is shown in Fig. 1.10.

Exercises 1.6

1.6.1 Show that an equable rhombus has height 4.

1.6.2 Find the dimensions of all equable rectangles.

1.6.3 If p is the perimeter and A is the area of an integer-sided rectangle, then what is the maximum value of p/A.

1.6.4 Suppose that an equable parallelogram is defined to have integer sides a and b and integer area, in which the ratio of the area to the perimeter is equal to 1 (that is, without the condition that there should be an integer distance between a pair of parallel sides). Show that one class of solutions is given by $a = u + v$ and $b = u - v$, where $v^2 \leqslant u(u - 4)$, and find another class of solutions in which $a + b$ is odd.

2 Circles and triangles

This chapter explores a variety of problems concerned with circles and triangles. A triangle ABC has a number of circles associated with it.

- There is the *circumcircle*, which is the circle passing through A, B, and C. The centre O is called the *circumcentre*; this is the point at which the perpendicular bisectors of the sides BC, CA, and AB concur. The radius of the circumcircle is denoted by R.
- There is the *incircle*, which is the circle touching the sides BC, CA, and AB, and which lies wholly within the triangle. The centre I is called the *incentre*; this is the point at which the three internal-angle bisectors meet. The radius of the incircle is denoted by r. Formulae connecting R, r, and the sides and area of ABC have already been given in eqns (1.3.1).
- The *nine-point circle*, passing through the feet of the altitudes, is not considered in this chapter, but appears in Section 8.6.
- Then there are the *escribed* circles or *excircles*, as they are sometimes called. The escribed circle opposite A touches BC, and the lines AB beyond B and AC beyond C. Its centre is denoted by I_1; this is the point where the internal bisector of angle A meets the external bisectors of angles B and C. The radius of this excircle is denoted by r_1. The escribed circles opposite B and C are similarly defined, with centres I_2 and I_3 and radii r_2 and r_3, respectively.

Some of the sections in this chapter are concerned with these circles that are associated with a triangle. Fig. 2.1 shows a triangle with its circumcircle and its incircle. There are also sections that are concerned with the special quadrilaterals associated with a circle, such as a *cyclic quadrilateral*, which is inscribed in a circle, so that the circle passes through its vertices, and an *inscribable quadrilateral*, which circumscribes a circle, so that the circle touches its sides and lies wholly within the quadrilateral. For problems concerning these circles a familiarity with the elementary circle theorems, such as the intersecting chord theorem, is assumed.

One of the main problems in this chapter is concerned with the *medians* of a triangle ABC. These are the lines AL, BM, and CN joining the vertices A, B, and C to the midpoints L, M, and N, respectively, of the opposite sides, and these lines concur at the centroid G. The question is whether an integer-sided triangle can have medians all of integer length. There are certainly an infinite number of solutions, and they appear to be relatively common, in the sense that there are about fifty solutions, with side lengths of up to about 5000.

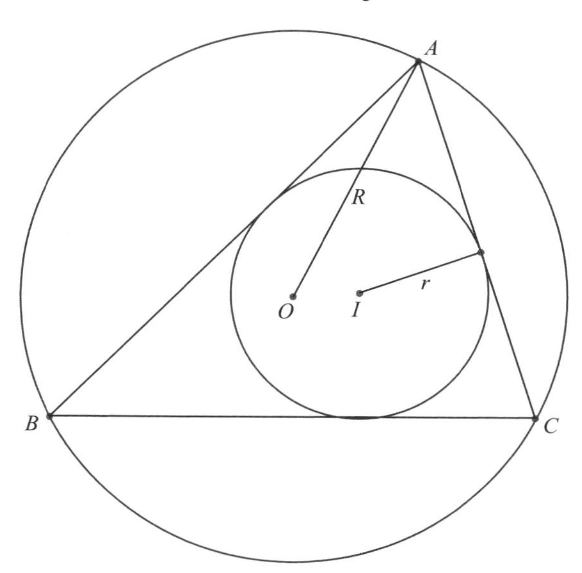

Fig. 2.1 The circumcircle and the incircle.

Finally, we present solutions to some problems concerning the existence of integer-sided triangles when the angles of the triangle bear simple relationships with one another, and concerning the number of integer-sided triangles with a given perimeter.

2.1 The circumradius R and the inradius r

R is the radius of the circumcircle of the integer-sided triangle ABC and r is the radius of the incircle. We recall the formulae relating R and r to the area $[ABC]$ and the side lengths a, b, and c. These are $R = abc/4[ABC]$ and $r = 2[ABC]/(a+b+c)$.

We assume throughout that a, b, and c are integral. So if $[ABC]$ is rational then R and r are rational. In fact, we restrict discussion to when $[ABC]$ is integral and investigate more precisely conditions for r and R to be integral, even though any Heron triangle may be enlarged to ensure that this happens.

Right-angled integer-sided triangles

R is half the length of the hypotenuse, since the circumcentre is at its midpoint. In a primitive triangle, since c is always odd, R is never an integer. When the sides are $2k(u^2 - v^2)$, $4kuv$, and $2k(u^2 + v^2)$, where k is any positive integer, then R is an integer and equal to $k(u^2 + v^2)$. Thus 5 is the smallest possible value of R for such triangles.

On the other hand, if triangle ABC is primitive (see Section 1.1) then

$$a + b + c = 2u(u + v) , \tag{2.1.1}$$
$$[ABC] = uv(u^2 - v^2) , \tag{2.1.2}$$

and, since $\frac{1}{2}r(a+b+c) = [ABC]$, we have $r = v(u - v)$ and r is always an integer. Also, r can be made to take on any positive integral value by a suitable choice for u and v.

Integral r for integer-sided triangles with integer area

Writing $a = \frac{1}{2}(m+n)$, $b = \frac{1}{2}(n+l)$, and $c = \frac{1}{2}(l+m)$, and using $rs = [ABC]$ and Heron's formula for $[ABC]$, we find that $lmn = 4r^2(l + m + n)$. Since l, m, and n have to be of the same parity for a, b, and c to be integral, they must therefore be even. Putting $l = 2u$, $m = 2v$, and $n = 2w$, we have $a = v + w$, $b = w + u$, $c = u + v$, and $uvw = r^2(u + v + w)$. This means that

$$u = \frac{r^2(v + w)}{vw - r^2} . \tag{2.1.3}$$

Supposing that u, v, and w are ordered with u greatest and w least, it follows that $u \geqslant \frac{1}{2}(v + w)$, and hence that $vw \leqslant 3r^2$. Also, $vw > r^2$.

For $r = 1$ there is only one solution with $v = 2$, $w = 1$, and $u = 3$. Hence the $(3, 4, 5)$ triangle is the only triangle with integer sides and integer area for which $r = 1$. Referring to Section 1.5 shows that, for triangles with integer sides and integer area, $r = 2$ if and only if the triangle is equable.

By inspection, there exists at least one triangle with integer sides and integer area for each integral value of r, given by

$$w = 1 , \qquad v = r^2 + 1 , \qquad u = r^2(r^2 + 2) , \tag{2.1.4}$$
$$a = r^2 + 2 , \quad b = r^4 + 2r^2 + 1 , \quad c = r^4 + 3r^2 + 1 . \tag{2.1.5}$$

Integral R for integer-sided triangles with integer area

Since $[ABC]$ is an integer, $R = abc/4[ABC]$, and $a/\sin A = b/\sin B = c/\sin C = 2R$, it follows that $\sin A$, $\sin B$, and $\sin C$ are all rational. From Section 1.1, integers p, q, u, and v exist with $\sin A = 2uv/(u^2 + v^2)$ and $\sin B = 2pq/(p^2 + q^2)$. Choosing

$$R = \frac{1}{4}(u^2 + v^2)(p^2 + q^2) \tag{2.1.6}$$

to remove denominators, we obtain the integer solution

$$a = uv(p^2 + q^2) , \tag{2.1.7}$$
$$b = pq(u^2 + v^2) , \tag{2.1.8}$$

Using $\sin C = \sin(A + B)$, we find that

$$c = uv(p^2 - q^2) + pq(u^2 - v^2)\,. \tag{2.1.9}$$

For R to be integral, either all of u, v, p, and q must be odd, or at least one of the pairs u and v or p and q must have both members even. It is possible that all of a, b, c, and R have a common factor, which may be extracted. On the other hand, the figure can be enlarged by multiplying all lengths by an integer scale factor. As an example, take $u = p = 3$ and $v = q = 1$, then $a = b = 30$, $c = 48$, and $R = 25$. If $p = q$ or $u = v$, then the triangle has a right angle.

Exercises 2.1

2.1.1 If ABC is a right-angled integer-sided triangle with rational R and r, then find the maximum value of h such that $R > hr$ for all such triangles.

2.1.2 Find all triangles having integer sides and integer area with $r = 3$.

2.1.3 If ABC is an integer-sided triangle with rational R and r, then find the maximum value of h such that $R > hr$ for all such triangles.

2.1.4 Prove that in no integer-sided right-angled triangle is it possible for $R = hr$, where h is an integer.

2.2 Intersecting chords and tangents

In this section we consider a circle and two chords AB and CD meeting at X. If X is an external point then there is a tangent TX to the circle touching it at T. The problem is to discover conditions so that AX, BX, CX, and DX are of integer length and also that TX is of integer length, when X is external. To make the problem more interesting we insist that the circle has radius R of integer length.

Intersection inside the circle

Let the chords AB and CD meet at X and let the diameter through X be EOF, where O is the centre of the circle and E and F lie on the circle. Write $AX = a$, $BX = b$, $CX = c$, $DX = d$, $OE = OF = R$, and $OX = h$. Then the intersecting chord theorem gives

$$ab = cd = (R + h)(R - h)\,. \tag{2.2.1}$$

Solutions exist with integers p, q, r, and s such that $a = pq$, $b = rs$, $c = pr$, $d = qs$, $R + h = ps$, and $R - h = qr$, from which $R = \frac{1}{2}(ps + qr)$ and $h = \frac{1}{2}(ps - qr)$. Since $R + h$ must be the greatest of the six factors, we must choose ps to be the largest product. Also, ps and qr must be of the same parity. For example, if we choose $p = 5$, $s = 4$, $q = 2$, and $r = 3$ then $R = 13$, $h = 7$, $a = 10$, $b = 12$, $c = 15$, and $d = 8$, the common product being 120, see Fig. 2.2. In general, other solutions exist by distributing the factors of $pqrs$ in $R + h$ and $R - h$ differently.

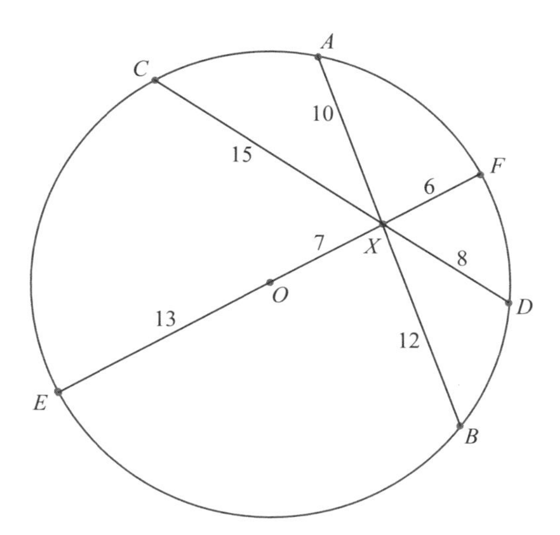

Fig. 2.2 The intersecting chord theorem.

Intersection outside the circle

Let $XEOF$ be a diametral chord cutting the circle at E and F, where O is the centre of the circle, and let XT be a tangent touching the circle at T. Let $XO = h$, $EO = OF = R$, and $XT = t$. Then the secant and tangent theorem gives

$$(h + R)(h - R) = t^2 = h^2 - R^2 . \qquad (2.2.2)$$

Now this is just the equation for a Pythagorean triple, so a parametric representation is $h = k(u^2 + v^2)$, $R = 2kuv$, and $t = k(u^2 - v^2)$, or $h = k(u^2 + v^2)$, $R = k(u^2 - v^2)$, and $t = 2kuv$, where u and v are positive coprime integers of opposite parity.

Suppose now that XAB is another chord through X meeting the circle at A and B, and let M be the midpoint of AB. Let $XA = a$, $XB = b$, $XM = s$, and $AM = MB = r$. Then the secant and tangent theorem gives

$$t^2 = ab = s^2 - r^2 , \quad \text{where} \quad a = s - r \quad \text{and} \quad b = s + r . \qquad (2.2.3)$$

It follows that integers m, p, and q exist so that $t = 2mpq$, $r = m(p^2 - q^2)$, $s = m(p^2 + q^2)$, $a = 2mq^2$, and $b = 2mp^2$, or $t = m(p^2 - q^2)$, $r = 2mpq$, $s = m(p^2 + q^2)$, $a = m(p - q)^2$, and $b = m(p + q)^2$. The constants k, u, v, m, p, and q must be chosen so that the value of t is the same in eqns (2.2.2) and (2.2.3), and so that $r < R$, since the length of the chord is less than the length of the diameter. The condition on t is algebraically the same condition as in the matching of the altitudes of the two component right-angled triangles forming a Heron triangle (see Section 1.3), so there is no need to give more than one example. We give the analogue of the $(13, 14, 15)$ triangle, where the altitude is 12, corresponding here to a value of $t = 12$. This choice involves the parameters $k = 3$, $u = 2$, $v = 1$, $p = 3$, $q = 2$, and $m = 1$.

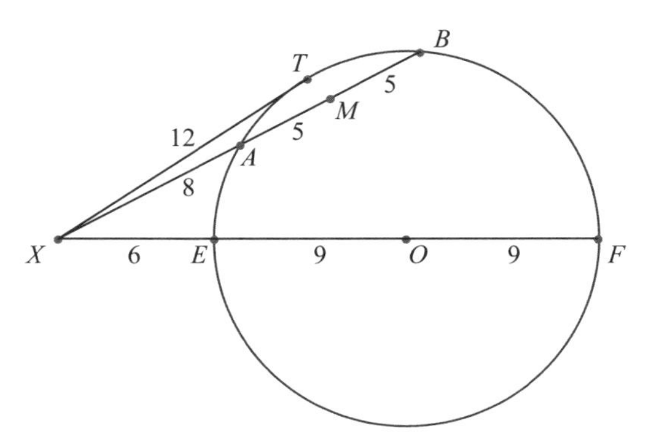

Fig. 2.3 The secant and tangent theorem.

Then $t = 2kuv = 2mpq = 12$, $h = 15$, $R = 9$, $r = 5$, $s = 13$, $a = 8$, and $b = 18$, see Fig. 2.3.

Exercise 2.2

2.2.1 The solution to Exercise 1.3.3 called for a list of all Heron triangles with altitude 40. Use the analogy pointed out in the last paragraph of Section 2.2 to find the possible values of t, h, and R for the corresponding circle and external point X, supposing that these lengths are integers.

2.3 Cyclic quadrilaterals and inscribable quadrilaterals

A cyclic quadrilateral is one that has a circle passing through all of its four vertices; the circle circumscribes the quadrilateral. An inscribable quadrilateral is one that has a circle touching all of its four sides; the circle is inscribed in the quadrilateral. A cyclic inscribable quadrilateral possesses both a circumscribing circle and an inscribed circle.

In this section we consider the following problems.

(i) For the cyclic quadrilateral, under what conditions does the quadrilateral have integer sides, integer diagonals, and a circumscribing circle of integer radius?

(ii) For the inscribable quadrilateral, under what conditions does the quadrilateral have integer sides and an inscribed circle of integer radius?

(iii) For the cyclic inscribable quadrilateral, under what conditions does the quadrilateral have integer sides and four other specified distances in the configuration also of integer length, but with no condition on the radius of either circle?

Cyclic quadrilaterals

Cyclic quadrilaterals of a particular kind, with integer sides and integer radius, are commonplace. For example, two copies of an integer-sided right-angled triangle, joined together with the common (even) hypotenuse as diameter, form a cyclic rectangle, which has integer diagonals as well.

Taking our cue from this example, we limit our investigation to cyclic quadrilaterals with integer sides and integer diagonals, and which lie in a circle of integer radius.

Let $ABCD$ be such a cyclic quadrilateral and suppose that $\angle ACB = \angle ADB = x$, $\angle ABD = \angle ACD = y$, $\angle CAB = \angle CDB = z$, and $\angle CAD = \angle CBD = w = 180^\circ - x - y - z$. The fact that triangles ADB and BCD have integer sides means that the cosines of x, y, z, and w are rational. Also, the equations $AB = 2R\sin x$, $AD = 2R\sin y$, $BC = 2R\sin z$, and $CD = 2R\sin w$ imply that the sines of x, y, z, and w are also rational. It follows from the theory of Section 1.1 that there exist integers u, v, s, t, p, and q such that

$$\begin{aligned} \sin x &= \frac{2uv}{u^2+v^2}, \quad \sin y = \frac{2st}{s^2+t^2}, \quad \sin z = \frac{2pq}{p^2+q^2}, \\ \cos x &= \frac{u^2-v^2}{u^2+v^2}, \quad \cos y = \frac{s^2-t^2}{s^2+t^2}, \quad \cos z = \frac{p^2-q^2}{p^2+q^2}. \end{aligned} \tag{2.3.1}$$

We can now ensure that the sides and diagonals are integers by choosing

$$R = \frac{1}{4}(u^2+v^2)(s^2+t^2)(p^2+q^2). \tag{2.3.2}$$

Note that the condition for R to be integral is that either both elements of two of the pairs u and v, s and t, and p and q are odd or both elements of one of the pairs is even. With the chosen value for R, all denominators disappear and, using the above equations, we obtain, after using trigonometrical formulae for expressions such as $\sin(y+z)$, the following integer values for the sides and diagonals:

$$\begin{aligned} AB &= uv(s^2+t^2)(p^2+q^2), \\ AD &= st(p^2+q^2)(u^2+v^2), \\ BC &= pq(u^2+v^2)(s^2+t^2), \\ AC &= uv(p^2-q^2)(s^2+t^2) + pq(u^2-v^2)(s^2+t^2), \\ BD &= uv(s^2-t^2)(p^2+q^2) + st(u^2-v^2)(p^2+q^2), \\ CD &= uv(s^2-t^2)(p^2-q^2) + pq(s^2-t^2)(u^2-v^2) \\ &\qquad + st(p^2-q^2)(u^2-v^2) - 4uvstpq. \end{aligned} \tag{2.3.3}$$

For example, with $u = 3$, $v = 1$, $s = 2$, $t = 1$, $p = 5$, and $q = 1$ we find, after cancelling a common factor of 5, that $R = 65$, $AB = 78$, $BC = 50$, $CD = 120$, $DA = 104$, $AC = 112$, and $BD = 130$. As a second example, with $u = 4$, $v = 2$, $p = 3$, $q = 2$, $s = 4$, and $t = 1$ we find that $R = 1105$, $AB = 1768$, $BC = 2040$,

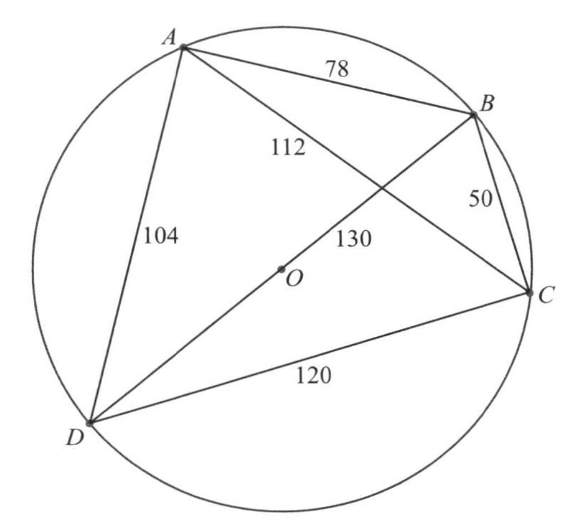

Fig. 2.4 A cyclic quadrilateral with integer sides, integer diagonals, and integer radius.

$CD = 1152$, $DA = 1040$, $AC = 1904$, and $BD = 2184$. The first of these examples is shown in Fig. 2.4.

Results may always be checked with the aid of Ptolemy's theorem, which states that a quadrilateral $ABCD$ is cyclic if and only if

$$AB \cdot CD + BC \cdot DA = AC \cdot BD\,. \qquad (2.3.4)$$

Inscribable quadrilaterals

These are defined to be quadrilaterals that possess an incircle. $ABCD$ is such a quadrilateral if and only if $AB + CD = BC + DA$. Then $AB = a + b$, $BC = b + c$, $CD = c + d$, and $DA = d + a$, where a, b, c, and d are the lengths of the tangents from A, B, C, and D to the incircle, respectively. Given any integer values of a, b, c, and d or half odd integer values of a, b, c, and d, it is always possible to draw a quadrilateral with these tangent lengths having an incircle, but its radius is not generally an integer. In order to create an interesting problem we therefore restrict our attention to inscribable quadrilaterals, which not only have integer sides but whose sides touch a circle of integer radius.

Let the centre of the incircle be O and suppose that AB, BC, CD, and DA touch the incircle at P, Q, R, and S, respectively. Denote $\angle AOP = x$, $\angle BOQ = y$, $\angle COR = z$, and $\angle DOS = w$. Then $\tan x = a/r$, $\tan y = b/r$, $\tan z = c/r$, and $\tan w = d/r$. Now $x + y + z + w = 180°$, so $\tan(x + y) = -\tan(z + w)$. Hence $r(a + b)/(r^2 - ab) + r(c + d)/(r^2 - cd) = 0$, from which

$$r^2 = \frac{bcd + acd + abd + bcd}{a + b + c + d}\,. \qquad (2.3.5)$$

This is a linear equation in d with solution

$$d = \frac{(a+b+c)r^2 - abc}{bc + ca + ab - r^2}. \qquad (2.3.6)$$

Any integer choice of a, b, c, and r making d positive provides an inscribable quadrilateral, with integer sides and integer radius, when enlarged by a scale factor equal to the denominator of d.

For example, with $a = 3$, $b = 4$, and $c = 2$ we obtain $d = (9r^2 - 24)/(26 - r^2)$. Possible solutions are when $r = 2$, 3, 4, and 5 with $d = 6/11$, $57/17$, 12, and 201, respectively. The four solutions are as follows:

(i) $a = 33$, $b = 44$, $c = 22$, $d = 6$, and $r = 22$;

(ii) $a = 51$, $b = 68$, $c = 34$, $d = 57$, and $r = 51$;

(iii) $a = 3$, $b = 4$, $c = 2$, $d = 12$, and $r = 4$; and

(iv) $a = 3$, $b = 4$, $c = 2$, $d = 201$, and $r = 5$.

The third of these cases is illustrated in Fig. 2.5.

Cyclic inscribable quadrilaterals

Let $ABCD$ be a cyclic inscribable quadrilateral, that is, a cyclic quadrilateral which also has an incircle. The problem we consider applies only to those cyclic inscribable quadrilaterals in which both pairs of opposite sides meet.

Suppose then that AB and DC meet at Q, and DA and CB meet at P. The problem we consider is how to ensure that all of the line segments PA, PB, AB, BC, CD, DA, QB, and QC of the quadrilateral are integers. Note that we make no requirement that either the radius of the circumcircle or the radius of the incircle should be integers, or that the diagonals AC, BD, and PQ should be integers.

Let $PA = s$, $PB = t$, $AB = x$, $BC = y$, $CD = z$, $DA = w$, $QB = u$, and $QC = v$, where these are all integers. It turns out that a solution can be obtained in terms of four positive integer parameters l, m, n, and p, in which each of s, t, u, v, w, x, y, and z are quartic expressions in terms of the parameters. A common multiplier may be included as an enlargement factor, and sometimes a common factor appears that may be divided out.

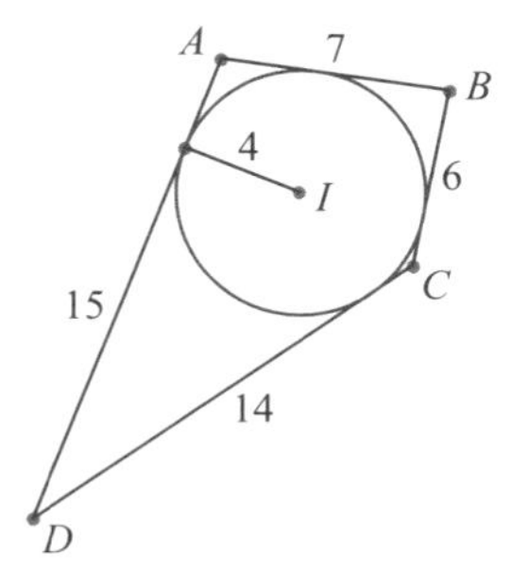

Fig. 2.5 An inscribable quadrilateral with integer sides and integer radius.

Since we require that $\boldsymbol{PA}\cdot\boldsymbol{PD}=\boldsymbol{PB}\cdot\boldsymbol{PC}$, positive integers l, m, n, and p must exist such that $\boldsymbol{PA}=lm\boldsymbol{i}$, $\boldsymbol{PD}=np\boldsymbol{i}$, $\boldsymbol{PB}=ln\boldsymbol{j}$, and $\boldsymbol{PC}=mp\boldsymbol{j}$, where $\boldsymbol{i}$ and $\boldsymbol{j}$ are unit vectors. Since AB and CD are not parallel, we must choose $m\neq n$. Also, since $PD>PA$ and $PC>PB$, we must choose $p>l$. Setting up oblique co-ordinates in this way, with P as origin, ensures that we have a general solution of the intersecting chord theorem

$$PA\cdot PD=PB\cdot PC\,, \tag{2.3.7}$$

and so, by the converse of the theorem, $ABCD$ is cyclic. Furthermore, the choices made ensure that both pairs of opposite sides meet.

Now $\boldsymbol{AB}=l(n\boldsymbol{j}-m\boldsymbol{i})$ and $\boldsymbol{DC}=p(m\boldsymbol{j}-n\boldsymbol{i})$. Since $|n\boldsymbol{j}-m\boldsymbol{i}|=|m\boldsymbol{j}-n\boldsymbol{i}|$, it follows that $lz=px$. Hence $x=kl$ and $z=kp$, where k may be rational (rather than integral). The condition that $ABCD$ should be inscribable is

$$x+z=w+y\,. \tag{2.3.8}$$

Now $w=np-lm$ and $y=mp-ln$, so the condition (2.3.8) becomes $k(p+l)=(m+n)(p-l)$. Thus $k=(m+n)(p-l)/(p+l)$, $x=(m+n)(p-l)l/(p+l)$, and $z=(m+n)(p-l)p/(p+l)$. Now, by Menelaus' theorem for triangle PCD, we have

$$\left|\frac{DQ}{QC}\frac{CB}{BP}\frac{PA}{AD}\right|=1\,. \tag{2.3.9}$$

After some algebra this gives $v=m(p-l)(mp-ln)/(p+l)(n-m)$. Similarly, Menelaus' theorem for triangle PAB gives

$$\left|\frac{PD}{DA}\frac{AQ}{QB}\frac{BC}{CP}\right|=1\,. \tag{2.3.10}$$

After some more algebra this gives $u=n(p-l)(mp-ln)/(n-m)(p+l)$. We now enlarge the figure by the factor $(n-m)(p+l)$ and (ignoring a further possible enlargement factor) we have finally

$$\begin{aligned}
s&=lm(n-m)(p+l)\,,\\
t&=ln(n-m)(p+l)\,,\\
x&=l(m+n)(p-l)(n-m)\,,\\
y&=(mp-ln)(n-m)(p+l)\,,\\
z&=p(m+n)(p-l)(n-m)\,,\\
w&=(np-lm)(n-m)(p+l)\,,\\
u&=n(p-l)(mp-ln)\,,\\
v&=m(p-l)(mp-ln)\,,
\end{aligned} \tag{2.3.11}$$

For example, with $l=1$, $m=10$, $n=11$, and $p=2$, after dividing out a common factor of 3, we obtain $s=10$, $t=11$, $x=7$, $y=9$, $z=14$, $w=12$, $u=33$, and $v=30$. As a second example, with $l=1$, $p=2$, $m=2$, and $n=3$ we obtain $s=6$, $t=9$, $x=5$, $y=3$, $z=10$, $w=12$, $u=3$, and $v=2$, see Fig. 2.6.

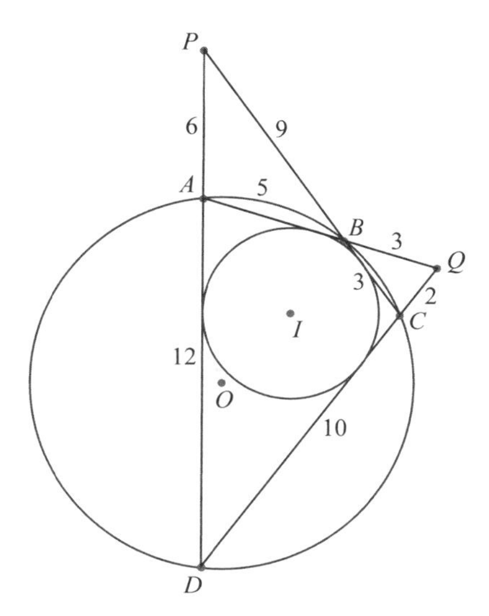

Fig. 2.6 A cyclic inscribable quadrilateral.

Exercises 2.3

2.3.1 Is it true or not that, if integers a, b, c, d, e, and f exist such that $ab+cd=ef$, then a cyclic quadrilateral necessarily exists with sides a, b, c, and d and diagonals e and f?

2.3.2 Show that, in a cyclic quadrilateral with $AB=a$, $BC=b$, $CD=c$, $DA=d$, $BD=e$, and $AC=f$, then $e^2=[(a^2+d^2)bc+(b^2+c^2)ad]/(bc+ad)$, with a similar expression for f^2.

2.3.3 A quadrilateral has sides $AB=1$, $BC=3$, $CD=4$, and $DA=2$. What is the radius of its incircle?

2.3.4 Is it possible to order the sides of a convex polygon of 2000 sides of lengths $1, 2, 3, \ldots, 2000$ in such a way that it possesses an incircle?

2.3.5 In the figure in which PDC is a triangle and ABQ is a transversal with A on PD, B on PC, and Q on the extension of DC, find AB and BC given that $ABCD$ is a cyclic inscribable quadrangle and $PA=40$, $PB=50$, $AD=35$, $DC=27$, and $CQ=8$.

2.4 The medians of a triangle

The *medians* of a triangle ABC are the lines from the vertices A, B, and C to the midpoints L, M, and N, respectively, of the opposite sides. If we write $AL=l$ then, by Apollonius' theorem,

$$4l^2 = 2b^2 + 2c^2 - a^2\,. \tag{2.4.1}$$

To prove this, complete the parallelogram $ABXC$ and use the cosine rule on triangle ABX, noting that $AX = 2AL = 2l$ and $\angle ABX = 180° - \angle BAC$. In what follows we write $a = 2A$, $b = 2B$, and $c = 2C$, and these symbols cannot be confused with the angles of the triangle because of the context in which they are used.

We consider in this section the conditions under which one or more medians may have integer length in an integer-sided triangle.

One integral median

It is very easy to choose a, b, and c so that l is an integer. We have $4l^2 = 2b^2 + 2c^2 - a^2$. Clearly, a must be even, so writing $a = 2A$, $b = 2B$, and $c = 2C$ we obtain

$$(B + C)^2 + (B - C)^2 = A^2 + l^2 .$$

The general solution of this derives from the famous identity

$$(ps + qr)^2 + (pr - qs)^2 = (pr + qs)^2 + (ps - qr)^2 \tag{2.4.2}$$

and leads to the parametric representation

$$\begin{aligned} a &= 2(pr + qs)\,, & b &= p(s + r) - q(s - r)\,, \\ c &= p(s - r) + q(s + r)\,, & l &= ps - qr\,. \end{aligned} \tag{2.4.3}$$

Any integer values of p, q, r, and s that lead to positive a, b, c, and l will do provided that the triangle inequalities are satisfied. For example, with $p = 3$, $q = 2$, $r = 1$, and $s = 4$ we obtain $a = 22$, $b = 9$, $c = 19$, and $l = 10$.

Two integral medians

In an isosceles triangle it is possible for all three sides and two medians to be integral. For example, $AB = BC = 14$, $CA = 12$, and $AL = CN = 11$. This is because, when $a = c$, in order to make AL and CN integers, there is only one equation to be satisfied. This is $l^2 = 2B^2 + A^2$ and this equation can be solved parametrically, see Exercise 2.4.1.

Three integral medians

The general analysis of the situation in which all three medians of an integer-sided triangle have integer length is similar to the problem where all of the face diagonals of an integer-sided rectangular parallelepiped have integer length. In both problems there are three symmetric expressions in the squares of a, b, and c that have to be made perfect squares. It is not surprising, therefore, that the situation is similar as regards their solution. In both cases there is a known two-parameter system of solutions (see eqns (1.4.2) and (2.4.9)), and there are further solutions, which may belong to other parametric systems of solutions, and there are probably additional sporadic solutions.

In the case of three integral medians the equations to be solved are

$$3A^2 + l^2 = 3B^2 + m^2 = 3C^2 + n^2 = 2(A^2 + B^2 + C^2)\,. \tag{2.4.4}$$

The example with lowest A, B, and C appears to be $A = 68$, $B = 85$, $C = 87$, $l = 158$, $m = 131$, and $n = 127$, see Fig. 2.7. All solutions are involutional, in the sense that the solution A, B, C, l, m, and n leads to a solution $A' = l$, $B' = n$, $C' = m$, $l' = 3A$, $m' = 3C$, and $n' = 3B$.

It is possible to express the solution of eqns (2.4.4) in terms of three rational numbers k, h, and $t = \tan \frac{1}{2}\theta$, together with a constraint. This solution is

$$\begin{aligned} A &= \frac{2[k(1-t^2)+2th]}{1+t^2}\,, \quad B = h - k\,, \quad C = h + k\,, \\ l &= \frac{2[h(1-t^2)-2tk]}{1+t^2}\,, \quad m = 3kh + 1\,, \quad n = 3kh - 1\,, \end{aligned} \tag{2.4.5}$$

together with the constraint

$$9k^2h^2 + 1 = \cos^2\theta(9k^2 + h^2) + 16hk\sin\theta\cos\theta + \sin^2\theta(k^2 + 9h^2)\,. \tag{2.4.6}$$

All this shows is that the three straightforward eqns (2.4.4) may be replaced by the single much more complicated eqn (2.4.6).

In order to obtain integer values one simply multiplies up by the least common multiple of the denominators. The solutions in Table 2.1 are the result of a computer search and are exhaustive for low values of A, B, and C.

If integer parameters are preferred then, with $k = y/e$, $h = x/d$, and $t = \tan \frac{1}{2}\theta$, we have

$$\begin{aligned} A &= 2(yd\cos\theta + xe\sin\theta)\,, \quad B = xe - yd\,, \quad C = xe + yd\,, \\ l &= 2(xe\cos\theta - yd\sin\theta)\,, \quad m = 3xy + ed\,, \quad n = 3xy - ed\,. \end{aligned} \tag{2.4.7}$$

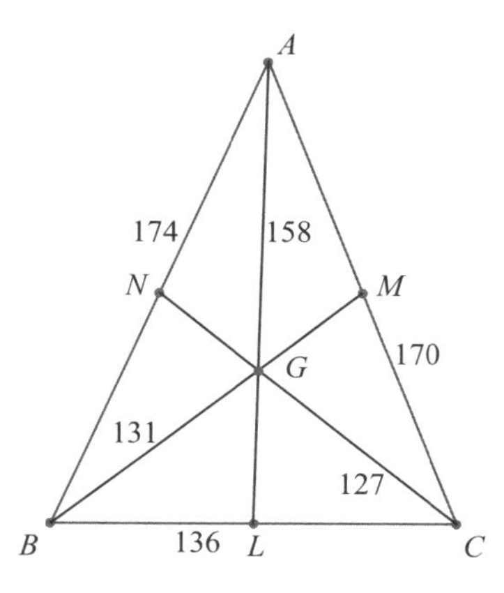

Fig. 2.7 A triangle with integer sides and integer medians.

Table 2.1 Integer-sided scalene triangles with $a = 2A$, $b = 2B$, and $c = 2C$ and integer medians l, m, and n.

k	h	t	A	B	C	l	m	n
1/2	43	1/5	68	85	87	158	131	127
2/3	43	1/3	158	127	131	204	261	255
31/33	16/3	7/13	328	145	207	142	529	463
11/31	16/3	1/26	142	463	529	984	621	435
5/6	89/39	4/15	290	113	243	244	523	367
2/5	89/39	1/20	244	367	523	870	729	339
11/12	61/3	35/58	442	233	255	208	683	659
4/11	61/3	2/29	208	659	683	1326	765	699
41/69	16/3	2/9	386	327	409	632	725	587
23/41	16/3	1/5	632	587	725	1158	1227	981
83/121	2	1/6	314	159	325	404	619	377
121/249	2	27/323	404	377	619	942	975	477
25/39	29/12	3/17	446	277	477	640	881	569
13/25	29/12	2/17	640	569	881	1338	1431	831
79/177	11/6	1/17	466	491	807	1252	1223	515
59/79	11/6	17/98	1252	515	1223	1398	2421	1473
163/279	8/3	4/25	774	581	907	1312	1583	1025
93/163	8/3	13/85	1312	1025	1583	2322	2721	1743
95/157	171/38	26/119	1524	1223	1603	2410	2879	2251
157/285	171/38	9/49	2410	2251	2879	4572	4809	3669
65/127	11/8	19/391	1306	877	1917	2680	3161	1129
127/195	11/8	3/38	2680	1129	3161	3918	5751	2631
143/241	7/2	9/47	1778	1401	1973	2924	3485	2521
241/429	7/2	14/81	2924	2521	3485	5334	5919	4203
11/12	229/111	17/59	1664	509	1323	1118	2963	2075
4/11	229/111	23/1003	1118	2075	2963	4992	3969	1527
109/173	8/5	11/105	1640	839	1929	2482	3481	1751
173/327	8/5	2/27	2482	1751	3481	4920	5787	2517
53/72	23/5	9/29	2312	1391	1921	2430	4017	3297
24/53	23/5	3/25	2430	3297	4017	6936	5763	4173
76/129	61/3	11/42	2704	2547	2699	4498	4765	4507
43/76	61/3	10/41	4498	4507	4765	8112	8097	7641
38/47	43	23/49	3158	1983	2059	2524	4949	4855
47/114	43	15/119	2524	4855	4949	9474	6177	5949
251/453	20/3	69/335	2848	2769	3271	5350	5473	4567
151/251	20/3	11/46	5350	4567	5473	8544	9813	8307
28/33	193/3	133/25	3514	2095	2151	2384	5437	5371
11/28	193/3	19/17	2384	5371	5437	10 542	6453	6285
11/12	389/123	20/49	2822	1105	2007	1592	4771	3787
4/11	389/123	1/28	1592	3787	4771	8466	6021	3315
67/183	47/30	11/714	1516	2197	3537	5690	4979	1319
61/67	47/30	35/187	5690	1319	4979	4548	10 611	6591

Table 2.1 continued

k	h	t	A	B	C	l	m	n
34/63	179/51	4/25	3444	3181	4337	6782	7157	5015
21/34	179/51	20/97	6782	5015	7157	10 332	13 011	9543
209/303	26/9	17/75	3396	1999	3253	4198	6343	4525
101/209	26/9	85/752	4198	4525	6343	10 188	9759	5997
19/54	169/87	1/74	1266	2491	3593	6052	4777	1645
18/19	169/87	2/7	6052	1645	4777	3798	10 779	7473
223/227	10	13/18	4450	2047	2493	1004	6917	6463
227/669	10	5/243	1004	6463	6917	13 350	7479	6141

The constraint becomes

$$e^2d^2 + 9x^2y^2 - x^2e^2 - y^2d^2 = 8(xe\sin\theta + yd\cos\theta)^2 . \tag{2.4.8}$$

In all cases θ is an acute angle and parameters have to be chosen to ensure positive values for A, B, C, l, m, and n.

Kevin Buzzard of Imperial College, London (private communication) informs me that the equations define a so-called 'elliptic surface of positive rank' and that as a consequence there will exist infinitely many parametric solutions, but almost certainly infinitely many sporadic solutions that lie on none of these parametric solutions.

Euler gave one parameterisation, namely

$$\begin{aligned} A &= 2u(9u^4 - 10u^2v^2 - 3v^4) , \\ B &= u(9u^4 + 26u^2v^2 + v^4) - v(9u^4 - 6u^2v^2 + v^4) , \\ C &= u(9u^4 + 26u^2v^2 + v^4) + v(9m^4 - 6u^2v^2 + v^4) . \end{aligned} \tag{2.4.9}$$

But, whilst $u = 2$ and $v = 1$ gives the case $A = 404$, $B = 377$, and $C = 619$, higher values of u and v give values of A, B, and C larger than any in our list. This illustrates the difficulty in this problem of obtaining an efficient parameterisation.

It is not known whether a triangle exists with integer sides, integer medians, and integer area. Adding one more equation to be satisfied produces a much more difficult situation, so if solutions exist then they are likely to be sparse.

Exercise 2.4

2.4.1 Find the general solution for the case of two integer medians of an isosceles integer-sided triangle.

2.5 The incircle and the excircles

In this section we show that a triangle with integer sides and integer inradius may be enlarged so that its area and the radii of its three excircles are integers. We use the notation that r is the radius of the incircle, s is the semi-perimeter of the triangle ABC, and r_1, r_2, and r_3 are the radii of the excircles opposite A, B, and C, respectively.

Theorem 2.5.1 *Suppose that ABC is a triangle with side lengths a, b, and c, $s = \frac{1}{2}(a+b+c)$ is its semi-perimeter, r is the inradius of the triangle ABC, and r_1, r_2, and r_3 are the radii of the excircles opposite A, B, and C, respectively. If r, a, b, and c are rational, then r_1, r_2, and r_3 are rational. In consequence, any triangle with integer sides and integer inradius may be enlarged so that all of r_1, r_2, and r_3 are integers.*

Proof If I is the incentre then, from the area relationship

$$[ABC] = [IBC] + [ICA] + [IAB]\,, \tag{2.5.1}$$

we have $[ABC] = rs$. Also, if I_1 is the excentre opposite A then we have

$$[ABC] = [I_1AB] + [I_1AC] - [I_1BC] = r_1(s-a)\,. \tag{2.5.2}$$

Similarly, $[ABC] = r_2(s-b) = r_3(s-c)$.

If r, a, b, and c are rational then, from the above equations, so is $[ABC]$, and hence so are r_1, r_2, and r_3. Multiplying by the lowest common multiple of the denominators of these rational numbers produces an enlargement in which the area $[ABC]$ and all of r_1, r_2, and r_3 are integers. □

We go further and deduce simple equations to determine r_1, r_2, and r_3. From Heron's formula for $[ABC]$ it is an immediate consequence that

$$r = \left[\frac{(s-a)(s-b)(s-c)}{s}\right]^{1/2} \tag{2.5.3}$$

and

$$r_1 = \left[\frac{s(s-b)(s-c)}{s-a}\right]^{1/2}, \tag{2.5.4}$$

with similar equations for r_2 and r_3. It follows that $rr_1 = (s-b)(s-c)$ or, on using the notation $a = v+w$, $b = w+u$, and $c = u+v$, we have $rr_1 = vw$, $rr_2 = wu$, and $rr_3 = uv$.

Recall from Section 2.1 that $r = 2$ for integer-sided triangles if and only if the triangle ABC is equable. There are just three such triangles with all of r_1, r_2, and r_3 integers. They are shown in Table 2.2. The second entry in the table is illustrated in Fig. 2.8.

Exercise 2.5

2.5.1 Find r, r_1, r_2, and r_3 when (i) $a = 12$, $b = 50$, and $c = 58$, and (ii) $a = 18$, $b = 20$, and $c = 34$.

Table 2.2 The three integer-sided triangles with $r = 2$ and r_1, r_2, and r_3 integers.

a	b	c	u	v	w	r	r_1	r_2	r_3
5	12	13	10	3	2	2	3	10	15
6	8	10	6	4	2	2	4	6	12
7	15	20	14	6	1	2	3	7	42

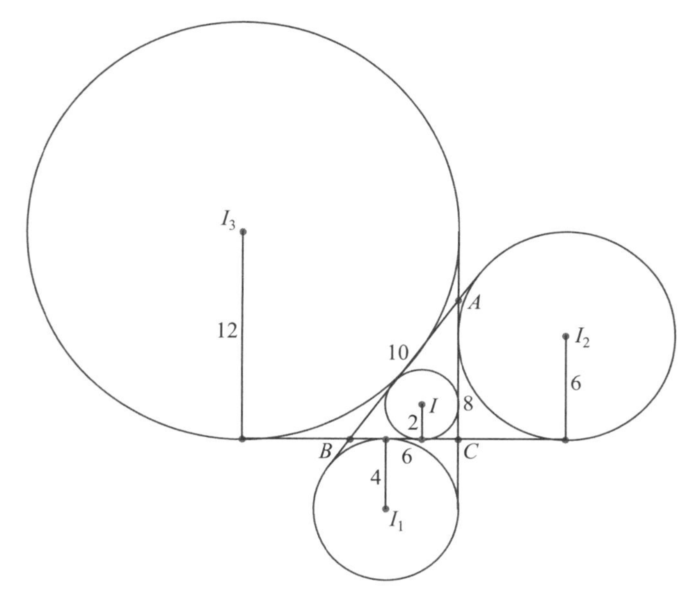

Fig. 2.8 A triangle with integer sides, integer inradius, and integer exradii.

2.6 The number of integer-sided triangles of given perimeter

We now consider a problem of a somewhat different kind, but still connected with integer-sided triangles. It is to do with counting and, as such, is a combinatorial problem. The problem is to determine the number of *non-congruent* integer-sided triangles having a specified integer perimeter p. It is possible to give a complete solution to this problem, though, since there are six different cases depending on the value of a certain parameter modulo 6, we do not provide every detail.

It is important to appreciate the significance of the phrase 'non-congruent' as far as the count is concerned, so we provide two illustrative examples. Suppose, as a first example, that $p = 12$. Then an enumeration of cases shows that there are three non-congruent triangles with sides $a = 5$, $b = 5$, and $c = 2$; $a = 5$, $b = 4$, and $c = 3$; and $a = 4$, $b = 4$, and $c = 4$. Other possibilities found by permuting the values of a, b, and c lead to triangles that are congruent to one or other of these three triangles and are therefore excluded from the count. Suppose, as a second example, that $p = 9$. Then an enumeration of cases shows that again there are three non-congruent triangles with sides $a = 4$, $b = 4$, and $c = 1$; $a = 4$, $b = 3$, and $c = 2$; and $a = 3$, $b = 3$, and $c = 3$. In other words, we want to count all possible selections of positive integers a, b, and c in which $a + b + c = p$, $a \geqslant b \geqslant c$, $b + c > a$, $c + a > b$, and $a + b > c$.

We suppose that a, b, and c are integers satisfying $a \geqslant b \geqslant c$ and, as in Section 2.1, let $a = \frac{1}{2}(m + n)$, $b = \frac{1}{2}(n + l)$, and $c = \frac{1}{2}(l + m)$, so that the perimeter $p = a + b + c = l + m + n$. The condition that a, b, and c are integers implies that l, m, and n are integers of the same parity—odd when p is odd and even when p is even. The condition that $a \geqslant b \geqslant c$ leads to the condition that $n \geqslant m \geqslant l$. The advantage of using l, m, and n rather than a, b, and c is that it converts the triangle inequalities $b + c > a$, $c + a > b$, and $a + b > c$ into the simpler relations $l, m, n > 0$.

The problem now reduces to counting, when p is even, the number of selections of positive even integers l, m, and n such that $l + m + n = p$, and, when p is odd, the number of selections of positive odd integers l, m, and n such that $l + m + n = p$, and where in both cases the choices of l, m, and n are limited by the condition that $n \geqslant m \geqslant l$. A further simplification can be made when p is even. Then l, m, and n are also even and we can write $p = 2N$, $l = 2u$, $m = 2v$, and $n = 2w$. Hence, for even perimeter $p = 2N$ the problem reduces to counting the number of selections of positive integers u, v, and w such that $u + v + w = N$ and $w \geqslant v \geqslant u$. The reader should check that in the case $p = 2N = 12$ the three possible selections for u, v, and w are $u = 1$, $v = 1$, and $w = 4$; $u = 1$, $v = 2$, and $w = 3$; and $u = 2$, $v = 2$, and $w = 2$, and in the case $p = 9$ the three possible selections for l, m, and n are $l = 1$, $m = 1$, and $n = 7$; $l = 1$, $m = 3$, and $n = 5$; and $l = 3$, $m = 3$, and $n = 3$.

Theorem 2.6.1 *For each value of the integer $N > 2$, the number of integer-sided triangles with perimeter $2N - 3$ is the same as the number of integer-sided triangles with perimeter $2N$.*

Proof If p is even, let $p = 2N$; then, from the text above, the problem for even perimeter reduces to counting the number of selections of positive integers u, v, and w such that $u + v + w = N$ and $w \geqslant v \geqslant u$. If p is odd and equal to $2N - 3$, then we require the number of selections of positive odd integers l, m, and n such that $l + m + n = 2N - 3$ and $n \geqslant m \geqslant l$. We claim that there is an obvious 1–1 correspondence between these counting problems defined by $u = \frac{1}{2}(l + 1)$, $v = \frac{1}{2}(m + 1)$, and $w = \frac{1}{2}(n + 1)$. For, if l, m, and n are odd integers satisfying $l + m + n = 2N - 3$, then u, v, and w are positive integers satisfying $u + v + w = \frac{1}{2}(l + m + n) + \frac{3}{2} = N$. And, if u, v, and w are positive integers satisfying $u + v + w = N$, then l, m, and n given by $l = 2u - 1$, $m = 2v - 1$, and $n = 2w - 1$ are odd positive integers satisfying $l + m + n = 2(u + v + w) - 3 = 2N - 3$. Also, $w \geqslant v \geqslant u \Leftrightarrow n \geqslant m \geqslant l$. □

If three real numbers x, y, and z are elements of a set $\{x, y, z\}$ then this set is called a *triple* and it does not matter in which order you write down the elements. On the other hand, if we define $[x, y, z] = \{\{x\}, \{x, y\}, \{x, y, z\}\}$, then $[x, y, z]$ is called an *ordered triple*. Clearly, two ordered triples $[x_1, y_1, z_1]$ and $[x_2, y_2, z_2]$ are equal if and only if $x_1 = x_2$, $y_1 = y_2$, and $z_1 = z_2$.

If $\{u, v, w\} = \{1, 2, 3\}$ then there is only one triple, but from the same integers you can form six ordered triples $[1, 2, 3]$, $[1, 3, 2]$, $[2, 3, 1]$, $[2, 1, 3]$, $[3, 1, 2]$, and $[3, 2, 1]$. Similarly, from the triple $\{1, 1, 4\}$ you can form three ordered triples $[4, 1, 1]$, $[1, 4, 1]$, and $[1, 1, 4]$, and from the triple $\{2, 2, 2\}$ there is just one ordered triple $[2, 2, 2]$. If you choose from the ordered triples those for which $w \geqslant v \geqslant u$ then they are in 1–1 correspondence with the triples. In our counting problem, by virtue of Theorem 2.6.1, the problem reduces to counting triples $\{u, v, w\}$, where u, v, and w are positive integers satisfying $u + v + w = q$, and where q is a positive integer. As we shall see below, in order to count the number of triples, it is necessary to keep track of the total number of ordered triples. Six separate cases need to be covered, as the working is slightly different depending on the value of q (mod 6).

- *Case 1,* $q = 6k$
We are looking for the number q_3 of partitions of q into three positive integers u, v, and w, in which we insist that $w \geqslant v \geqslant u$. The reason for this, as explained above, is that a choice such as $u = 1$, $v = 2$, and $w = 3$ leads to a triangle with the same side lengths as choices such as $u = 2$, $v = 1$, and $w = 3$, so that we include in q_3 only one of the six possibilities with the integers 1, 2, and 3. However, as the working below shows, in order to determine q_3 we also have to keep a count of the number Q_3 of all sets of values of u, v, and w, so in Q_3 all six possibilities with the integers 1, 2, and 3 would be included. In other words, q_3 counts triples and Q_3 counts ordered triples.
When $u = v = w$ there is only one possibility with the common value $2k$. This leads to a contribution of 1 to q_3 and a contribution of 1 to Q_3.
When $u = v$, with w different, then u can range from 1 to $3k - 1$, except that we have already counted $u = 2k$. So there are $3k - 2$ possibilities of this kind. This provides a contribution of $3k - 2$ to q_3 and a contribution of $9k - 6$ to Q_3.
Now,
$$Q_3 = {}^{q-1}C_2 = \frac{1}{2}(6k-1)(6k-2) = 18k^2 - 9k + 1\,, \tag{2.6.1}$$
so the contribution to Q_3 of choices with distinct values of u, v, and w is $(18k^2 - 9k + 1) - (9k - 6) - 1 = 18k^2 - 18k + 6$. Thus the contribution to q_3 of choices with distinct values of u, v, and w is $3k^2 - 3k + 1$. It follows that
$$q_3 = 1 + (3k-2) + (3k^2 - 3k + 1) = 3k^2 = \frac{q^2}{12}\,. \tag{2.6.2}$$

- *Case 2,* $q = 6k + 1$
The contribution to q_3 and Q_3 when $u = v = w$ is 0. The contribution to q_3 when $u = v$ and w is distinct is $3k$, and the contribution to Q_3 is $9k$. Now, $Q_3 = \frac{1}{2}[6k(6k-1)] = 18k^2 - 3k$. The contribution to Q_3 when u, v, and w are distinct is $18k^2 - 12k$. So the contribution to q_3 in this case is $3k^2 - 2k$. It follows that $q_3 = 3k^2 + k = (q^2 - 1)/12$.

We leave the reader to supply details of the following results.

- *Case 3*, $q = 6k + 2$
  Here $q_3 = (q^2 - 4)/12$.
- *Case 4*, $q = 6k + 3$
  Here $q_3 = (q^2 + 3)/12$.
- *Case 5*, $q = 6k + 4$
  Here $q_3 = (q^2 - 4)/12$.
- *Case 6*, $q = 6k + 5$
  Here $q_3 = (q^2 - 1)/12$.

For example, when $p = 2005$, Theorem 2.6.1 tells us that the number of distinct triangles is the same as for $p = 2008$. Then $q = 1004 = 6k + 2$, with $k = 167$. The number of triangles is $(q^2 - 4)/12 = 84\,001$.

Exercise 2.6

2.6.1 Find the number of distinct integer-sided triangles with perimeter 1001.

2.7 Triangles with angles u, $2u$, and $180° - 3u$

In this section we consider triangles whose angles are u, $2u$, and $180° - 3u$, and obtain formulae for the side lengths when these are integers. The problem is also treated in Shklarsky *et al.* (1993). Each such triangle is similar to one in which the bisector of the angle of size $2u$ is of integer length.

The general solution

Let $\angle ABC = u$ and $\angle BCA = 2u$; then $\angle CAB = 180° - 3u$. From the sine rule we have

$$\begin{aligned} a &= 2R\sin 3u = 2R(3\sin u - 4\sin^3 u)\,, \\ b &= 2R\sin u\,, \\ c &= 2R\sin 2u = 4R\sin u\cos u\,. \end{aligned} \tag{2.7.1}$$

It follows that $c^2 = b^2 + ab = 16R^2\sin^2 u(1 - \sin^2 u)$. Assuming that a, b, and c are integers with highest common factor k, it follows that b and $a + b$ are k times perfect squares, and so coprime integers p and q exist, with $p > q$, such that

$$a = k(p^2 - q^2)\,, \quad b = kq^2\,, \quad c = kpq\,. \tag{2.7.2}$$

For example, with $p = 4$, $q = 3$, and $k = 1$ we obtain $a = 7$, $b = 9$, and $c = 12$.

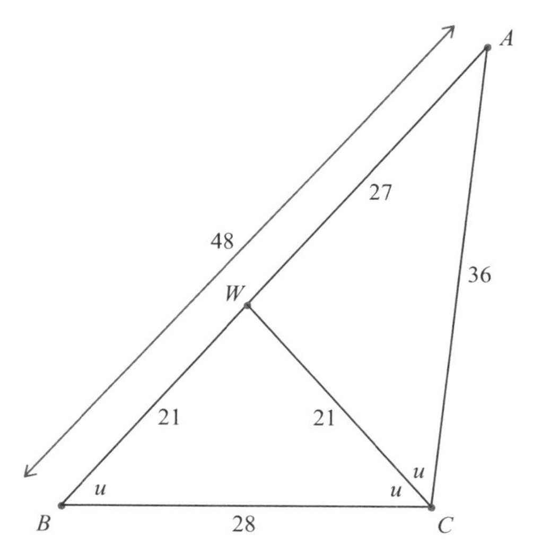

Fig. 2.9 A triangle with an internal bisector of integer length.

Internal angle bisector

If the above triangle is magnified by a factor of 4 so that $a = 28$, $b = 36$, and $c = 48$ and the internal bisector CW of angle BCA is drawn, then since $AW/BW = AC/BC = 9/7$ (see Theorem A.6.1) it follows that $AW = 27$ and $BW = 21$. Now triangle WBC is isosceles with $CW = BW = 21$. This example is illustrated in Fig. 2.9. In this way we can always construct a triangle in which an angle bisector is of integer length. We do not, at present, consider the more difficult problem of when all three internal bisectors are of integer length. This problem is considered in Section 2.8. It can be proved by applications of the cosine rule that

$$CW = \frac{[ab(a+b-c)(a+b+c)]^{1/2}}{a+b}. \tag{2.7.3}$$

It may be checked with the above values of a, b, and c that CW comes to 21.

Exercise 2.7

2.7.1 Use $p = 5$ and $q = 2$ with an appropriate value of k to find another triangle with an internal bisector of $\angle BCA$ that is of integral length.

2.8 Integer r and integer internal bisectors

Given the triangle ABC, let U, V, and W be the intersections of the internal bisectors with BC, CA, and AB, respectively. We now show how to find integer-sided triangles in which the inradius r and the lengths of all three internal angle bisectors AU, BV, and CW are integers.

The clue to the possibility of r and the lengths of all internal bisectors being integers arises from the formulae

$$AI = r \operatorname{cosec} \frac{1}{2}A\,, \quad BI = r \operatorname{cosec} \frac{1}{2}B\,, \quad CI = r \operatorname{cosec} \frac{1}{2}C\,. \tag{2.8.1}$$

It follows that, if we can find triangles in which r and all of $\sin \frac{1}{2}A$, $\sin \frac{1}{2}B$, and $\sin \frac{1}{2}C$ are rational, then AI, BI, and CI are all rational. Now, if AU, BV, and CW are the internal bisectors then we have

$$AU = \frac{(a+b+c)AI}{b+c}\,, \quad BV = \frac{(a+b+c)BI}{c+a}\,, \quad CW = \frac{(a+b+c)CI}{a+b}\,, \tag{2.8.2}$$

and hence these lengths are also rational. Finally, an enlargement of the triangle by the least common multiple of the denominators of r, AU, BV, and CW produces integer values for all of them.

We do not attempt to obtain general formulae for a, b, c, r, AU, BV, and CW in terms of parameters, as the variety of possibilities and the difficulty of the equations involved makes any such attempt unreasonable, if not impossible. However, the way to construct all triangles with the required properties is not difficult, and we explain why this is so and give an example to illustrate the method.

In a Heron triangle ABC the cosines and sines of A, B, and C are rational. However, if we wish to ensure that the cosines and sines of $\frac{1}{2}A$, $\frac{1}{2}B$, and $\frac{1}{2}C$ are rational, then we can only ensure this if we choose the component triangles of the Heron triangle in a particular way. They must both arise from Pythagorean triples whose parameters u and v are themselves the short legs of a Pythagorean triple, and then the triangles have to be matched up suitably. An example is the best way of illustrating what happens.

The values of u and v chosen are the short legs 4 and 3, and 12 and 5. Take $u = 4$ and $v = 3$; then from Table 1.1 we get the triple $(7, 24, 25)$. Take $u = 12$ and $v = 5$; then we get the triple $(119, 120, 169)$. We can match these by enlarging the first by a factor of 5 so that the altitude from A has length 120. We then have a Heron triangle in which $c = AB = 125$, $b = AC = 169$, and $a = BC = 35 + 119 = 154$. Now, $\cos B = 7/25$, so that $\sin \frac{1}{2}B = 3/5$ and $\cos \frac{1}{2}B = 4/5$. Likewise, $\sin \frac{1}{2}C = 5/13$ and $\cos \frac{1}{2}C = 12/13$. Then $\sin \frac{1}{2}A = \cos \frac{1}{2}(B + C) = 48/65 - 15/65 = 33/65$, so that $\cos \frac{1}{2}A = 56/65$. Now the semi-perimeter $s = (169 + 125 + 154)/2 = 448/2 = 224$ and $[ABC] = 60 \times 154 = 9240$. It follows, from $rs = [ABC]$, that $r = 9240/224 = 165/4$. Hence $AI = r \operatorname{cosec} \frac{1}{2}A = 325/4$, $BI = r \operatorname{cosec} \frac{1}{2}B = 275/4$, and $CI = r \operatorname{cosec} \frac{1}{2}C = 429/4$. Then, from eqns (2.8.2), $AU = 448/294 \times 325/4 = 2600/21$, $BV = 448/279 \times 275/4 = 30800/279$, and $CW = 448/323 \times 429/4 = 48048/323$, see Fig. 2.10. With an enlargement by a factor of $4 \times 21 \times 323 \times 93 = 2\,523\,276$ we obtain a Heron triangle with integer r and integer internal bisectors.

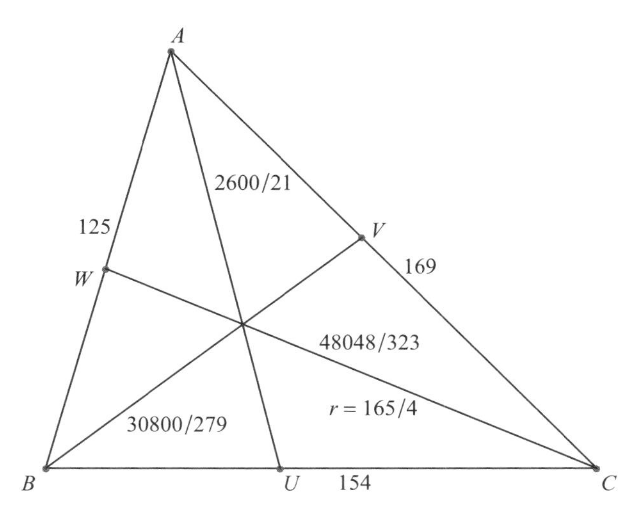

Fig. 2.10 A triangle with integer sides and rational angle bisectors.

Exercises 2.8

2.8.1 In the example given above, are the external bisectors of the angles also of rational length?

2.8.2 What goes wrong with the calculation of the internal angle bisectors if we match up the $(119, 120, 169)$ and $(119, 408, 425)$ triangles to form a Heron triangle?

2.9 Triangles with angles u, nu, and $180° - (n+1)u$

The case $n = 2$ was analysed in Section 2.7. In this section we generalise those results. We write $\angle A = u$, $\angle B = nu$, and $\angle C = 180° - (n+1)u$. For integer-sided triangles with these angles, for which the sides have no common factor, it is possible to express the sides in terms of two parameters. The reason for this is easy to see. By the sine rule we have $b/a = \sin nu/\sin u$ and $c/a = \sin(n+1)u/\sin u$.

Now, in an integer-sided triangle, the cosine of each angle is rational, so we may take $2\cos u = p/q$, where p and q are coprime integers. Next observe that, for any integer value of n, $\sin nu/\sin u$ is expressible as a polynomial in $\cos u$ of degree $n-1$ (by De Moivre's theorem). A parameterisation in terms of p and q may therefore be obtained as follows: from above, b/a is a homogenous polynomial of degree $n-1$ in p/q, and c/a is a homogeneous polynomial of degree n in p/q. We may therefore set $a = q^n$ and both b and c then become homogeneous polynomials of degree n in p and q. We now give the results for $n = 3$ and $n = 4$ and leave the reader to fill in the details of the calculation.

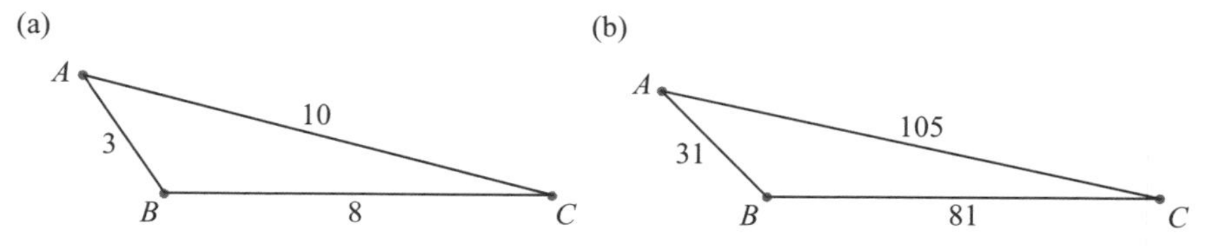

Fig. 2.11 An integer-sided triangle with (a) $\angle B = 3\angle A$, and (b) $\angle B = 4\angle A$.

Degree $n = 3$

We have $\sin 3u/\sin u = 4\cos^2 u - 1$ and $\sin 4u/\sin u = 4\cos u(2\cos^2 u - 1)$. The required parameterisation is therefore

$$a = q^3\,, \quad b = q(p^2 - q^2)\,, \quad c = p(p^2 - 2q^2)\,. \tag{2.9.1}$$

Since $0^\circ < u < 45^\circ$, we must have $2q > p > q\sqrt{2}$. The triangle with these angles having least perimeter comes from the values $p = 3$ and $q = 2$, and has sides $a = 8$, $b = 10$, and $c = 3$, see Fig. 2.11(a).

Degree $n = 4$

We have $\sin 4u/\sin u = 4\cos u(2\cos^2 u - 1)$ and $\sin 5u/\sin u = 16\cos^4 u - 12\cos^2 u + 1$. The required parameterisation is therefore

$$a = q^4\,, \quad b = pq(p^2 - 2q^2)\,, \quad c = p^4 - 3p^2q^2 + q^4\,. \tag{2.9.2}$$

Since $0^\circ < u < 36^\circ$, we have $2q > p > \frac{1}{2}q(1 + \sqrt{5})$. The triangle with these angles having least perimeter comes from the values $p = 5$ and $q = 3$, and has sides $a = 81$, $b = 105$, and $c = 31$, see Fig. 2.11(b).

Exercises 2.9

2.9.1 Find two triangles with angles u, $3u$, and $180^\circ - 4u$ that have one side equal to 48.

2.9.2 Find the parametric representation for the sides in the cases $n = 5$ and $n = 6$, and the triangles of least perimeter corresponding to these cases.

3 Lattices

In number theory work on lattices is of fundamental importance, and has been applied to problems as diverse as the rational approximation of real numbers and the representation of integers as the sum of four squares. In this chapter we do little more than scratch the surface. In Section 3.1 we give some basic definitions and theorems about two-dimensional lattices. Section 3.2 is devoted to an account of Pick's theorem, concerning the number of lattice points within or on the boundary of a polygon whose vertices are at lattice points. Finally, in Section 3.3 we show how to solve the problem of determining which lattice points lie on straight lines with equations of the form $lx + my = n$, where l, m, and n are integers, and then when l, m, and n are restricted to be positive integers.

3.1 Lattices and the square lattice

We consider lattices in the vector space $\mathbb{R}^2$, endowed with the usual scalar product. This is such that, if $\boldsymbol{a}$ and $\boldsymbol{b}$ are vectors with rectangular Cartesian components (a_1, a_2) and (b_1, b_2), respectively, then $\boldsymbol{a} \cdot \boldsymbol{b} = a_1 b_1 + a_2 b_2$.

A *lattice in* $\mathbb{R}^2$ is defined to be a set

$$\Lambda = \{n_1 \boldsymbol{e}_1 + n_2 \boldsymbol{e}_2 \mid n_1, n_2 \in \mathbb{Z}\}, \tag{3.1.1}$$

where $(\boldsymbol{e}_1, \boldsymbol{e}_2)$ forms a basis for $\mathbb{R}^2$.

The members of the set determine the *lattice points*.

We give two examples.

Example 3.1.1 Let $\boldsymbol{e}_1 = \boldsymbol{i}$ and $\boldsymbol{e}_2 = \boldsymbol{j}$, where $\boldsymbol{i}$ and $\boldsymbol{j}$ are unit vectors in the positive x- and y-directions, respectively. Then Λ is the *fundamental square lattice* and the lattice points are the elements of $\mathbb{Z}^2$, which may be represented by rectangular Cartesian co-ordinates (n_1, n_2), where n_1 and n_2 are integers. In what follows we denote the fundamental square lattice by L.

Example 3.1.2 Let $\boldsymbol{e}_1 = (1, 0)$ and $\boldsymbol{e}_2 = (\frac{1}{2}, \frac{1}{2}\sqrt{3})$. Then Λ is the *hexagonal lattice*.

In crystallography, given a lattice it is possible to define a *reciprocal lattice*. This is the set $\Lambda_R = \{n_1 \boldsymbol{g}_1 + n_2 \boldsymbol{g}_2 \mid n_1, n_2 \in \mathbb{Z}\}$, where the vectors $\boldsymbol{g}_1$ and $\boldsymbol{g}_2$ satisfy $\boldsymbol{g}_i \cdot \boldsymbol{e}_j = \delta_{ij}$. In Example 3.1.2, $\boldsymbol{g}_1 = (1, -1/\sqrt{3})$ and $\boldsymbol{g}_2 = (0, 2/\sqrt{3})$. The reciprocal lattice is a hexagonal lattice, but the basis vectors are of a different length to those in Λ. In the square lattice of Example 3.1.1, the reciprocal lattice vectors are of the same length and in the same direction as those in Λ.

It is important to appreciate that a change of basis in the definition of Λ may produce a lattice that determines the same lattice points. Obviously, the lattices in Examples 3.1.1 and 3.1.2 lead to distinct sets of lattice points. Let us consider some more examples.

Consider the lattice $\Lambda = \{n_1 \boldsymbol{f}_1 + n_2 \boldsymbol{f}_2 \,|\, n_1, n_2 \in \mathbb{Z}\}$, where $\boldsymbol{f}_1 = 2\boldsymbol{i}$ and $\boldsymbol{f}_2 = \boldsymbol{j}$. This is clearly not the same as the fundamental square lattice because the lattice points with an odd x co-ordinate are missing. It is, in fact, a rectangular lattice with one side twice the length of the other.

On the other hand, consider the lattice $\Lambda = \{n_1 \boldsymbol{f}_1 + n_2 \boldsymbol{f}_2 \,|\, n_1, n_2 \in \mathbb{Z}\}$, where $\boldsymbol{f}_1 = 2\boldsymbol{i} + \boldsymbol{j}$ and $\boldsymbol{f}_2 = \boldsymbol{i} + \boldsymbol{j}$. We claim that this is the same as the fundamental square lattice. The reason for this is, firstly, that every lattice point of Λ has integer co-ordinates of the form $(2n_1 + n_2, n_1 + n_2)$ and so they are also lattice points of the fundamental square lattice. Secondly, if (N_1, N_2) are the integer co-ordinates of any lattice point of the fundamental square lattice, then it is the same as the lattice point of Λ with $n_1 = N_1 - N_2$ and $n_2 = 2N_2 - N_1$. It seems appropriate to say that lattices defined with respect to different bases, but which determine the same set of lattice points, are *equivalent*.

Note that any lattice point may be chosen as the origin and properties of the lattice are independent of this choice. This is because the environment of each lattice point is the same.

The set $F = \{\alpha_1 \boldsymbol{e}_1 + \alpha_2 \boldsymbol{e}_2 \,|\, 0 \leqslant \alpha_1, \alpha_2 < 1\}$ is called a *fundamental region* of Λ (or in crystallography a *primitive unit cell* of Λ) and, in general, it has the shape of a parallelogram. In the fundamental square lattice of Example 3.1.1 the fundamental region has, of course, the shape of a square. In the hexagonal lattice of Example 3.1.2 the fundamental region has the shape of a rhombus with an acute angle of $60°$. Since lattices defined with different bases sometimes determine identical sets of lattice points, it is clear that equivalent lattices may have different fundamental regions. However, the fundamental regions of equivalent lattices have one thing in common, they all have the same area.

We state without proof the fact that every vector $\boldsymbol{v} \in \mathbb{R}^2$ is such that there is a unique vector $\boldsymbol{u} \in F$ such that $\boldsymbol{v} - \boldsymbol{u} \in \Lambda$. This means that the sets $F + \lambda$, for $\lambda \in \Lambda$, tessellate $\mathbb{R}^2$.

Theorem 3.1.3 *Two lattices are equivalent (determine the same lattice points) if and only if the bases to which they are referred are related by a 2×2 unimodular matrix (an integer matrix with determinant equal to ± 1).*

Proof Suppose that $\Lambda = \{n_1 \boldsymbol{e}_1 + n_2 \boldsymbol{e}_2 \,|\, n_1, n_2 \in \mathbb{Z}\}$ and $\Lambda' = \{m_1 \boldsymbol{f}_1 + m_2 \boldsymbol{f}_2 \,|\, m_1, m_2 \in \mathbb{Z}\}$, and that

$$\begin{aligned} \boldsymbol{f}_1 &= a\boldsymbol{e}_1 + b\boldsymbol{e}_2\,, \\ \boldsymbol{f}_2 &= c\boldsymbol{e}_1 + d\boldsymbol{e}_2\,, \end{aligned} \tag{3.1.2}$$

where a, b, c, and d are integers satisfying the unimodular condition $D = ad - bc = \pm 1$. It is clear that any lattice point (m_1, m_2) of Λ' is the same as the lattice point $(am_1 + cm_2, bm_1 + dm_2)$ of Λ. The inverse transformation is

$$\begin{aligned} \boldsymbol{e}_1 &= \frac{1}{D}\left(d\boldsymbol{f}_1 - b\boldsymbol{f}_2\right), \\ \boldsymbol{e}_2 &= \frac{1}{D}\left(-c\boldsymbol{f}_1 + a\boldsymbol{f}_2\right), \end{aligned} \tag{3.1.3}$$

and, since $D = \pm 1$, it is clear that any lattice point (n_1, n_2) of Λ is the same as the lattice point $((dn_1 - cn_2)/D, (-bn_1 + an_2)/D)$ of Λ'. It follows that Λ and Λ' are equivalent. Conversely, if Λ and Λ' are equivalent then, since $(n_1, n_2) = (0, 1)$ corresponds to the lattice point $(-c/D, a/D)$ in Λ', it must be the case that $D|a$ and $D|c$. Similarly, $D|b$ and $D|d$, and hence $D^2|(ad - bc)$, that is, $D^2|D$. Thus $D = \pm 1$ and the transformation is unimodular. □

Theorem 3.1.4 *The areas of fundamental regions of two equivalent lattices Λ and Λ' are equal.*

Proof If Λ is given by eqn (3.1.1) then the area of the fundamental region F is given by $|\boldsymbol{e}_1 \times \boldsymbol{e}_2|$. If the bases $(\boldsymbol{e}_1, \boldsymbol{e}_2)$ and $(\boldsymbol{f}_1, \boldsymbol{f}_2)$ are related by eqns (3.1.2), where a, b, c, and d are integers such that $|ad - bc| = 1$, then the area of the fundamental region F' of Λ' is given by $|\boldsymbol{f}_1 \times \boldsymbol{f}_2| = |(a\boldsymbol{e}_1 + b\boldsymbol{e}_2) \times (c\boldsymbol{e}_1 + d\boldsymbol{e}_2)| = |(ad - bc)(\boldsymbol{e}_1 \times \boldsymbol{e}_2)| = |\boldsymbol{e}_1 \times \boldsymbol{e}_2|$. □

In the remainder of this section we consider only the fundamental square lattice L. From Example 3.1.1 we have $L = \{n_1\boldsymbol{i} + n_2\boldsymbol{j} \,|\, n_1, n_2 \in \mathbb{Z}\}$, and we suppose that A and B are lattice points of L with $(n_1, n_2) = (a, b)$ and (c, d), respectively. Now consider the lattice L' defined by $L' = \{m_1\boldsymbol{OA} + m_2\boldsymbol{OB} \,|\, m_1, m_2 \in \mathbb{Z}\}$. By Theorems 3.1.3 and 3.1.4, L and L' are equivalent if and only if the area of the triangle OAB is equal to one-half of the area of the unit square, that is, $[OAB] = \frac{1}{2}|ad - bc| = \frac{1}{2}$. We have proved the following theorem.

Theorem 3.1.5 *Let $\boldsymbol{OA}$ and $\boldsymbol{OB}$ belong to the fundamental square lattice L. A necessary and sufficient condition that the lattice L' based on $\boldsymbol{OA}$ and $\boldsymbol{OB}$ should be equivalent to L is that the area of the triangle OAB is $\frac{1}{2}$.* □

A point A of L is *visible* (from O) if there is no point of L on OA between O and A. Clearly, in order for (x, y) to be visible, it is necessary and sufficient for x and y to be coprime. If the highest common factor of x and y is h, then the number of lattice points between O and A (not counting O or A) is $h - 1$.

Theorem 3.1.6 *Suppose that A and B are points of L visible from O, and let $[OAB]$ be the area of the triangle T defined by OA and OB. If $[OAB] = \frac{1}{2}$ then there is no point of L inside T; and if $[OAB] > \frac{1}{2}$ then there is at least one point of L inside T or lying on AB.* □

Theorem 3.1.6 follows immediately from Theorem 3.1.5.

The following three simple examples illustrate Theorem 3.1.6. Firstly, define the points $A(1, 1)$ and $B(1, 2)$. Then $\boldsymbol{OA}$ and $\boldsymbol{OB}$ form a basis for a lattice equivalent to L, since $ad - bc = 1$ and $[OAB] = \frac{1}{2}$. Secondly, define the points $A(1, -1)$ and $B(1, 1)$; then $ad - bc = 2$ and $[OAB] = 1$, and a lattice point $(1, 0)$ lies on AB.

Thirdly, define the points $A(2,1)$ and $B(1,2)$; then $ad - bc = 3$ and $[OAB] = 3/2$, and a lattice point $(1,1)$ lies internal to T.

In this book we do not develop the theory of lattices beyond this point. However, it is worth mentioning Minkowski's theorem because it has important consequences in number theory. For example, it may be used to prove the famous theorem that every prime of the form 1 (mod 4) may be expressed uniquely as the sum of two perfect squares. Minkowski's theorem, as stated below, may of course be generalised to higher dimensions.

Theorem 3.1.7 (Minkowski) *Let S be a centrally-symmetric convex set in $\mathbb{R}^2$ such that the area of S is greater than four times that of the fundamental unit square. Then S contains within it at least one lattice point of the fundamental square lattice L.* □

Exercises 3.1

3.1.1 Determine the number of internal points, the number of boundary points, and the area of the triangle OAB when A has co-ordinates $(6,0)$ and B has co-ordinates $(6,4)$.

3.1.2 Determine the number of internal points, the number of boundary points, and the area of the triangle OAB when A has co-ordinates $(6,0)$ and B has co-ordinates $(3,4)$.

3.1.3 Determine the number of internal points, the number of boundary points, and the area of the triangle OAB when A has co-ordinates $(6,3)$ and B has co-ordinates $(6,4)$.

3.1.4 Determine the number of internal points, the number of boundary points, and the area of the triangle OAB when A has co-ordinates $(6,1)$ and B has co-ordinates $(4,4)$.

3.1.5 Determine the number of internal points, the number of boundary points, and the area of the triangle OAB when A has co-ordinates $(4,2)$ and B has co-ordinates $(6,4)$.

3.1.6 Estimate the number of lattice points lying inside the circle with equation $x^2 + y^2 = 20\,000$.

3.2 Pick's theorem

In this section we give a proof of a theorem, called Pick's theorem, which provides a formula for the area of a polygon whose vertices are situated at the points of the fundamental square lattice. The simplicity of the statement of the theorem means that it is understood and enjoyed (but seldom proved) by students at quite an early age. The proof we give is rather elaborate, but has the merit of being elementary. It treats the triangle in great detail, and then uses an obvious induction from triangles to polygons with four or more sides. Preferably, the reader should complete Exercises 3.1 before studying the proof of the theorem.

Theorem 3.2.1 (Pick) *The area F of a polygon, whose vertices are situated at lattice points of the fundamental square lattice, whose sides do not cross, and which has I internal lattice points and B boundary lattice points, including vertices, is given by $F = I + \frac{1}{2}B - 1$. (The fundamental squares are of area 1.)*

Before embarking on a proof of Pick's theorem we produce some evidence for supposing that such a result is valid.

Firstly, check the results of Exercises 3.1. The five triangles covered in Exercises 3.1.1 to 3.1 5 are representative of all possible shapes of triangle that can be drawn on the square lattice. Secondly, the formula is certainly correct for a triangle of area $\frac{1}{2}$, by Theorem 3.1.5, since then $I = 0$ and $B = 3$. Thirdly, if there is a formula at all then we should expect it to be linear in I and B, since each lattice point is, loosely speaking, surrounded by a fixed amount of area. Fourthly, if the formula is true for a given triangle, then it has to be true when that triangle is enlarged by an integer factor k; that is, the formula for the area must give k^2 times the initial amount. So let us consider a triangle in which, apart from the vertices, there are x, y, and z lattice points on the sides and i internal lattice points. Then the proposed formula for the area gives

$$F_1 = i + \frac{1}{2}(x + y + z + 3) - 1 = i + \frac{1}{2}(x + y + z) + \frac{1}{2}\,. \qquad (3.2.1)$$

After enlargement by a factor of k about any vertex, we find that

$$I = k^2 i + \frac{1}{2}k(k-1)(x + y + z) + \frac{1}{2}(k-1)(k-2)\,, \qquad (3.2.2)$$

$$B = \frac{1}{2}(kx + ky + kz + 3k) - 1\,. \qquad (3.2.3)$$

The derivation of these formulae for I and B is left as an exercise. It is now easily checked that the formula $F_k = I + \frac{1}{2}B - 1$ gives $F_k = k^2 F_1$, as hoped.

We now give a proof of Theorem 3.2.1.

Proof of Theorem 3.2.1 In this proof, when we refer to the number of boundary points on a line segment, this number is always stated **exclusive** of the end-points. We use the notation $[OABC]$ for the area of $OABC$ and $[XYZ]$ for the area of the triangle XYZ.

We establish the result for the five types of triangle that can exist on a lattice, by inscribing them inside a rectangle, for which the formula is first proved to be correct. Larson (1983) covers the first two comparatively simple cases. For completeness, we give a proof for all six cases, see Fig. 3.1.

- *Case 1*
  A rectangle $OABC$ has a boundary points on OA and b boundary points on AB. We have $F = ab + \frac{1}{2}(2a + 2b + 4) - 1 = ab + a + b + 1 = (a+1)(b+1) = [OABC]$.

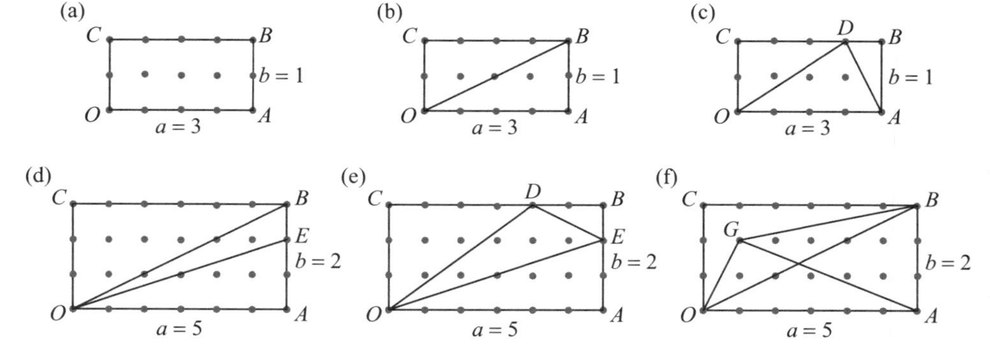

Fig. 3.1 Pick's theorem: $F = I + \frac{1}{2}B - 1$.

(a) Case 1, a rectangle $OABC$: $I = ab = 3$, $B = 2a + 2b + 4 = 12$, and $F = 8$.

(b) Case 2, a right-angled triangle OAB: $c = 1$, $j = 1$, $I = 1$, $B = 8$, and $F = 4$.

(c) Case 3, a triangle OAD: $c = 2$, $i = 2$, $k = 1$, $I = 2$, $B = 6$, and $F = 4$.

(d) Case 4, a triangle OEB: $y = 2$, $v = 1$, $i = 4$, $j = 1$, $m = 2$, $p = 0$, $q = 1$, $I = 1$, $B = 6$, and $F = 3$.

(e) Case 5, a triangle OED: $c = 3$, $d = 1$, $p = 0$, $q = 1$, $x = u = l = 0$, $v = 1$, $k = 3$, $i = 4$, $m = 2$, $I = 4$, $B = 4$, and $F = 5$.

(f) Case 6, a triangle OGB: $y = 2$, $r = s = t = 0$, $i = 4$, $j = 3$, $k = 4$, $l = 2$, $I = 3$, $B = 5$, and $F = 9/2$.

- *Case 2*
  With the notation as in Case 1, consider a right-angled triangle OAB with short legs parallel to the axes. Suppose that there are c points on the line OB and j points internal to OAB. Then, by symmetry, we have $2j + c = ab$.
  The formula gives $F = j + \frac{1}{2}(a + b + c + 3) - 1 = \frac{1}{2}(ab + a + b + 1) = \frac{1}{2}(a + 1)(b + 1) = [OAB]$.

- *Case 3*
  Again, with the notation as in Case 1, consider the triangle OAD, where D lies on BC. This is a triangle whose base is one of the sides of the rectangle. Suppose that there are c points on CD, d points on DB, x points on OD, y points on AD, i points inside OAD, j points inside ABD, and k points inside OCD. Then $i+j+k+x+y = ab$ and $c+d = a-1$. From Case 1 we have $[OABC] = ab+a+b+1 = i+j+k+x+y+a+b+1$. From Case 2 we have $[OCD] = k+\frac{1}{2}(b+c+x+1)$ and $[ABD] = j + \frac{1}{2}(b + d + y + 1)$.
  By subtraction, we have $[OAD] = i + \frac{1}{2}(x + y + 2a - c - d) = F = i + \frac{1}{2}(x + y + a + 3) - 1$, since $a = c + d + 1$.

- *Case 4*
  Again, with the notation as in Case 1, consider the triangle OEB with E on AB.

This is a triangle having a side which is part of one of the sides of the rectangle. Suppose that there are y points on OB, v points on OE, p points on BE, q points on EA, i points inside OBC, j points inside OEB, and m points inside OAE. We have $ab = i + j + v + y + m$ and $b = p + q + 1$. From Case 1 we have $[OABC] = ab + a + b + 1 = i + j + v + y + m + a + p + q + 2$. From Case 2 we have $[OAE] = m + \frac{1}{2}(v + q + a + 1)$ and $[OBC] = i + \frac{1}{2}(a + b + y + 1) = i + \frac{1}{2}(a + p + q + y + 2)$.
By subtraction, we find that $[OEB] = F = j + \frac{1}{2}(v + p + y + 3) - 1$.

- *Case 5*
  Again, with the notation as in Case 1, consider a triangle OED with vertices E and D on the sides AB and BC, respectively. Suppose that CD has c points, DB has d points, BE has p points, EA has q points, OD has x points, DE has u points, and OE has v points. Also suppose that there are k internal points in OCD, i internal points in ODE, l internal points in DBE, and m internal points in OAE. We have $a = c + d + 1, b = p + q + 1$, and $ab = i + k + l + m + u + v + x$. Then from Case 1 we have $[OABC] = i + k + l + m + u + v + x + c + d + p + q + 3$. From Case 2 we have $[OCD] = k + \frac{1}{2}(p + q + x + c + 2)$, $[OAE] = m + \frac{1}{2}(v + q + c + d + 2)$, and $[EBD] = l + \frac{1}{2}(u + p + d + 1)$.
  By subtraction, we have $[OED] = F = i + \frac{1}{2}(u + v + x + 1)$.
- *Case 6*
  Again, with the notation as in Case 1, consider a triangle OGB that has vertices at opposite corners of the rectangle and an internal vertex at G. We suppose, without loss of generality, that G lies within the triangle OBC. Suppose that OB has y points, GB has r points, GA has t points, and OG has s points. Also suppose that there are i points in OBC, j points in OGB, k points in AGB other than any in OGB, and l points in OGA other than any in OGB. Then $ab = i + k + l + t$. From Case 1 we have $[OABC] = i + k + l + t + a + b + 1$. From Cases 2 and 3 we have $[OBC] = i + \frac{1}{2}(a + b + y + 1)$ and $[OGA] + [AGB] = l + \frac{1}{2}(a + s + t + 1) + j + k + \frac{1}{2}(b + r + t + 1)$. From the equation $[AGB] + [OGA] + [OBC] = [OABC] + [OGB]$ we find that $[OGB] = F = j + \frac{1}{2}(y + r + s + 1)$.

We have now shown that the formula is true for all triangles, and the final stage, in moving from triangles to all polygons, is an induction on the number of sides of the polygon, supposing that the result is true for all polygons having sides up to and including k sides, where $k \geqslant 3$. First note that a polygon, even if re-entrant, has an interior diagonal. This diagonal cuts the polygon of $k+1$ sides into two polygons with no more than k sides, for each of which the theorem is true. It is now a simple exercise, which we leave to the reader, to complete the induction. You have to suppose that there are d lattice points on the diagonal, excluding the end-points, but d eventually cancels out. The proof is given in Larson (1983). □

Exercise 3.2

3.2.1 Find the area of the polygon with vertices at the following points: $(-3,4)$, $(-1,4)$, $(1,3)$, $(3,3)$, $(4,1)$, $(3,-1)$, $(1,-2)$, $(-2,-3)$, $(-4,-2)$, $(-5,0)$, and $(-4,3)$.

3.3 Integer points on straight lines

An *integer point* on a straight line or curve is a point whose rectangular Cartesian co-ordinates are integers. A *rational point* on a straight line or curve is a point whose rectangular Cartesian co-ordinates are rational numbers. Sometimes integer points are called lattice points, the presumption being that one is working with a square lattice, the sides of the square being of unit length.

The following theorem is given in all good undergraduate algebra textbooks.

Theorem 3.3.1 *If a and b are positive coprime integers then there exist integers x and y such that $xa + yb = 1$.* □

If one of a or b is negative then the theorem still holds, by changing the sign of x or y, as appropriate. The proof of this theorem is constructive, using the Euclidean algorithm, so we restrict ourselves to an example.

We find integers x and y that lie on the line with equation

$$74x + 51y = 1\,. \tag{3.3.1}$$

We have, from the Euclidean algorithm,

$$\begin{aligned}74 &= 1 \times 51 + 23\,,\\ 51 &= 2 \times 23 + 5\,,\\ 23 &= 4 \times 5 + 3\,,\\ 5 &= 1 \times 3 + 2\,,\\ 3 &= 1 \times 2 + 1\,.\end{aligned}$$

Now, working backwards, we find that $1 = 3 - 1 \times 2 = 3 - 1 \times (5 - 1 \times 3) = 2 \times 3 - 1 \times 5 = 2 \times (23 - 4 \times 5) - 1 \times 5 = 2 \times 23 - 9 \times 5 = 2 \times 23 - 9 \times (51 - 2 \times 23) = 20 \times 23 - 9 \times 51 = 20 \times (74 - 1 \times 51) - 9 \times 51 = 20 \times 74 - 29 \times 51$. From this we see that the point with co-ordinates $x = 20$ and $y = -29$ lies on the line given in eqn (3.3.1). Now, if (x, y) and (x', y') are two integer points lying on the line, then by subtraction $74(x - x') + 51(y - y') = 0$, so there exists an integer k such that $x' - x = -51k$ and $y' - y = 74k$. The general solution is therefore

$$x' = 20 - 51k\,, \quad y' = -29 + 74k\,. \tag{3.3.2}$$

As k runs through all integers, positive, zero, and negative, we obtain all integer points on the line.

If we want the integer points on the line with equation $74x + 51y = z$, where z is an integer, then the general solution is $x' = 20z - 51k$ and $y' = -29z + 74k$, where k is an integer.

Note that some lines with rational slope do not contain any integer points at all. For example, the line $2x + 6y = 3$ has no integer points on it because if x and y are integers then the left-hand side is even and the right-hand side is odd.

A considerably more difficult problem is to find those lines of negative slope that have non-negative integer solutions for x and y. The following theorem by Sylvester covers all possibilities. The proof is an abbreviation of the argument in the lengthy investigation entitled 'The postage stamp problem' by Gardiner (1987). The postage stamp problem is as follows: if you have an unlimited amount of stamps worth a pence and b pence, then which amounts can be made and which cannot be made using these stamps alone?

Theorem 3.3.2 *If a and b are positive coprime integers then the largest positive integer N which* ***cannot*** *be expressed in the form $N = xa + yb$, where x and y are both non-negative integers, is given by $N = ab - a - b$.*

Proof Without loss of generality we may take $a > b > 0$.

Lemma 3.3.3 *If m is a non-negative integer lying between 0 and $ab-a-b$ inclusive, and $n = ab-a-b-m$, then not both of m and n are expressible in the form $xa+yb$, where x and y are non-negative integers.*

Proof Suppose, in fact, that $m = xa + yb$ and $n = pa + qb$, where x, y, p, and q are non-negative integers. Clearly, we may take x and p to lie between 0 and $b-1$ inclusive, since $0 \leqslant m, n < ab$. Adding these two equations gives $m + n = ab - a - b = (x + p)a + (y + q)b$. That is,

$$ab = (x + p + 1)a + (y + q + 1)b\,. \tag{3.3.3}$$

Since a and b are coprime it follows that b divides $x + p + 1$. However, we know that this is less than or equal to $2b - 1$. Hence $b = x + p + 1$, and dividing eqn (3.3.3) by b one obtains $y + q + 1 = 0$. Thus one of y or q is negative and the other is non-negative. This proves the lemma and shows that at most one of m and n is expressible in the required form. □

Lemma 3.3.4 *If m is a non-negative integer lying between 0 and ab inclusive, then the number of values of m that are expressible in the form $xa + yb$, where x and y are non-negative integers, is $M = \frac{1}{2}(a + 1)(b + 1)$.*

Proof On the fundamental square lattice draw the co-ordinate axes and the line with equation $ax + by = ab$ meeting the x-axis at $B(b, 0)$ and the y-axis at $A(0, a)$. First note that, for fixed m in the interval $0 \leqslant m < ab$ that is expressible in the form $xa + yb = m$, with x and y non-negative integers, this can be done in exactly one way. For, if $m = x'a + y'b$ as well, with x' and y' non-negative integers, then $(x - x')a + (y - y')b = 0$. Now a and b are coprime, so $a|(y - y')$. However,

$0 \leqslant y, y' \leqslant a-1$ and so a cannot divide $y - y'$. The contradiction shows that $y = y'$ and $x = x'$. On the other hand, $m = ab$ is expressible in two ways represented by the lattice points A and B (but not by more, because, as a and b are coprime, the line AB passes through no other lattice point). By counting lattice points inside and on the boundary of the triangle OAB (but not counting both A and B), we have $M = \frac{1}{2}(a+1)(b+1)$. □

We are now in a position to prove Theorem 3.3.2.

Proof of Theorem 3.3.2 continued First, by Lemma 3.3.3, when m lies between 0 and $ab - a - b$ inclusive there are no more than $\frac{1}{2}(ab - a - b + 1)$ values of m that are expressible in the form $xa + yb = m$, with x and y non-negative integers. So, by Lemma 3.3.4, no less than $\frac{1}{2}(a+1)(b+1) - \frac{1}{2}(ab - a - b + 1) = a + b$ values of m between $ab - a - b + 1$ and ab inclusive are expressible. However, that is all of them. So, in fact, the number expressible between 0 and $N = ab - a - b$ inclusive is precisely half of them. By Lemma 3.3.3, we now see that, if $0 \leqslant m \leqslant N$ and m is expressible then $N - m$ is not expressible, and if m is not expressible then $N - m$ is expressible.

Now $1, 2, 3, \ldots, b-1$ are clearly not expressible, so $N-1, N-2, N-3, \ldots, N-b+1$ are expressible. Adding 1 to the value of y means that $N+b-1, N+b-2, N+b-3, \ldots, N+1$ are expressible. So also is $N + b = ab - a$, with $x = b - 1$ and $y = 0$. Starting with $N + 1$, we now have a sequence of b integers that are all expressible, and by increasing the value of y by 1 we see that the next b integers are expressible, and so on. The conclusion is that $N = ab - a - b$ is the largest integer that is not expressible. □

As an example of Theorem 3.3.2, if $a = 5$ and $b = 4$, then 0 is expressible and 11 is not, 1 is not but 10 is, 2 is not but 9 is, 3 is not but 8 is, 4 is and 7 is not, 5 is and 6 is not, and all positive integers greater than 11 are expressible.

Exercises 3.3

3.3.1 Find all pairs of integers x and y such that $35x + 12y = 1$.

3.3.2 Explain why it is not possible to find integers x and y such that $133x + 84y = 1$.

3.3.3 What is the smallest positive integer h for which integers x and y can be found such that $133x + 84y = h$?

3.3.4 For which values of N does the equation $7x + 4y = N$ have solutions for which x and y are non-negative integers?

4 Rational points on curves

This chapter is concerned with the identification of integer points and rational points on curves of degree two and three. Elliptic curves have been studied extensively throughout the twentieth century and they are at the centre of a deep and profound theory. My intention is to provide a reader new to such ideas with a readable account of the elementary theory, and at the same time to give enough examples to give a student a working knowledge of the subject. It is necessary, therefore, to explain how to classify singularities on curves, and to explain why it is necessary to embed curves in the projective plane.

4.1 Integer points on a planar curve of degree two

A *rational point* is defined to be a point on a curve whose rectangular Cartesian co-ordinates are both rational. There is a standard method of determining all rational points on a curve of second degree given one such point. Sometimes it is possible to determine the points on the curve with integer co-ordinates directly from the rational points, but there are cases in which alternative methods are required. We first consider the method of obtaining rational points by giving a number of illustrative examples.

Example 4.1.1 Consider the ellipse $E : x^2 + 2y^2 = 1$.

We use the fact that the point $(1, 0)$ lies on E. If (x, y) is another rational point on E then the gradient of the chord joining $(1, 0)$ to (x, y) is rational. Conversely, if m is rational then the line with equation $y = m(x - 1)$ passes through $(1, 0)$ and meets E where $x^2 + 2m^2(x - 1)^2 = 1$. Since this quadratic equation has one root $x = 1$, the other root is easily calculated to be $x = (2m^2 - 1)/(2m^2 + 1)$. This is rational and (x, y) lies on E, where $y = -2m/(2m^2 + 1)$, a value which is also rational. Putting $m = -u/v$, where u and v are coprime integers, we obtain the rational points

$$(x, y) = \left(\frac{2u^2 - v^2}{2u^2 + v^2}, \frac{2uv}{2u^2 + v^2} \right). \tag{4.1.1}$$

As the above argument implies, these are all the rational points on E.

Furthermore, if (x, y) is an integer point on the curve $x^2 + 2y^2 = z^2$, for some integer value of z, then (x, y, z) is proportional to the triple $(2u^2 - v^2, 2uv, 2u^2 + v^2)$. The only integer points on the curve E with equation $x^2 + 2y^2 = 1$ are obviously $(1, 0)$ and $(-1, 0)$, see Fig. 4.1.

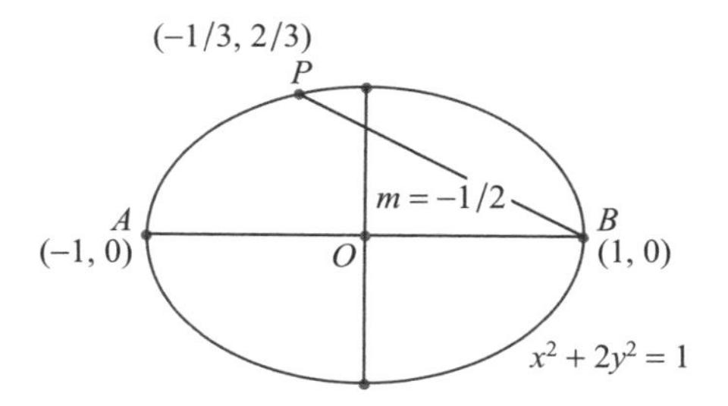

Fig. 4.1 A rational point on an ellipse.

Example 4.1.2 Consider the hyperbola $H : x^2 - 2y^2 = 1$.

We use the fact that $(1, 0)$ lies on H and consider the intersection of the line with equation $y = m(x-1)$ and H, where m is rational. This happens for values of x given by $x^2 - 2m^2(x-1)^2 = 1$. This quadratic equation has roots $x = 1$, corresponding to $y = 0$, and $x = (2m^2+1)/(2m^2-1)$, corresponding to $y = 2m/(2m^2-1)$. Putting $m = u/v$, where u and v are coprime integers, we find the rational points on the hyperbola to be $(1, 0)$ and

$$(x, y) = \left(\frac{2u^2 + v^2}{2u^2 - v^2}, \frac{2uv}{2u^2 - v^2}\right). \tag{4.1.2}$$

An attempt to find the integer points on the hyperbola from the rational points fails, as the denominators in eqn (4.1.2) are equal to ± 1 only if we can find integers u and v such that $2u^2 - v^2 = \pm 1$, which is essentially the same problem that we started with. We can make some progress by inspection. For example, $u = 1$ and $v = 1$ gives $(x, y) = (3, 2)$. Figure 4.2 shows this integer point and the others obtained from it by symmetry. We can then put $u = 2$ and $v = 3$ to obtain $(x, y) = (17, 12)$, but finding a few low-valued solutions is not going to help unless they form the basis of an induction argument.

In fact, the equation $x^2 - 2y^2 = 1$ belongs to the class of equations

$$x^2 - Ny^2 = 1\,, \tag{4.1.3}$$

where N is a positive integer but not a perfect square, which carries the name of Pell. It turns out that there is a method of obtaining all integer solutions for x and y for any such value of N. In fact, it was Lagrange who first proved that eqn (4.1.3) has infinitely many solutions, and the class of equations bears Pell's name by mistake. We illustrate the method for the case $N = 2$.

The method starts with the observation that $3 + 2\sqrt{2}$ is the smallest number of the form $x + y\sqrt{2}$, with x and y being positive integers, such that

$$x^2 - 2y^2 = 1\,. \tag{4.1.4}$$

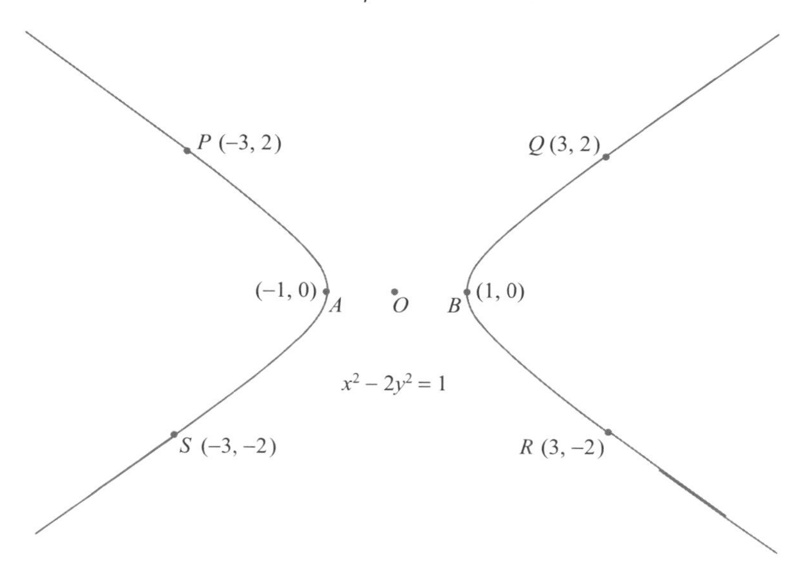

Fig. 4.2 Integer points on a hyperbola.

Now form $(3+2\sqrt{2})(x+y\sqrt{2})$ for any integer solution (x, y) of eqn (4.1.4), starting with $(3, 2)$. It equals $(3x+4y)+(2x+3y)\sqrt{2}$. If we now write

$$\begin{aligned} x' &= 3x + 4y\,, \\ y' &= 2x + 3y\,, \end{aligned} \tag{4.1.5}$$

then we find that $x'^2 - 2y'^2 = 9x^2 + 24xy + 16y^2 - 8x^2 - 24xy - 18y^2 = x^2 - 2y^2 = 1$. In this way, from one solution (x, y) we have generated another solution (x', y').

Theorem 4.1.3 *(x, y) is a solution of the equation $x^2 - 2y^2 = 1$ in positive integers x and y if and only if $x + y\sqrt{2} = (3+2\sqrt{2})^k$ for some positive integer k.*

Proof We have already shown that any (x, y) obtained from the above equation with a positive integer value of k provides a solution. It is a matter, therefore, of showing that there are no other solutions. Suppose, in fact, that there is a positive solution (c, d), not in the above sequence of solutions; then there must exist a value of k such that $(3+2\sqrt{2})^k < c+d\sqrt{2} < (3+2\sqrt{2})^{k+1}$. Multiplying by $(3+2\sqrt{2})^{-k} = (3-2\sqrt{2})^k$ we obtain $1 < (c+d\sqrt{2})(3-2\sqrt{2})^k < 3+2\sqrt{2}$. Defining integers x and y by $x+y\sqrt{2} = (c+d\sqrt{2})(3-2\sqrt{2})^k$, it is easy to check (by a similar analysis to that following eqn (4.1.5)) that $x^2 - 2y^2 = 1$. Also, $1 < x+y\sqrt{2} < 3+2\sqrt{2}$. It follows that $0 < 1/(x+y\sqrt{2}) < 1$, that is, $0 < x - y\sqrt{2} < 1$. Hence $2x > 1$, so x, being an integer, is greater than or equal to 1. However, $x - y\sqrt{2} < 1$, so $y > 0$. Hence $x+y\sqrt{2}$ is a positive solution of eqn (4.1.4) which is less than $3+2\sqrt{2}$. This is a contradiction to the fact that $(3, 2)$ is the least such solution. Hence the suggested solution (c, d) does not exist. □

A discussion of the Pell equation and how to obtain all solutions of eqn (4.1.3) is found in most books on number theory, see, for example, Silverman (1997). The proof that there is always one positive solution for any given non-square value of N is more difficult than showing that all solutions come from iteration. Some authors, such as Niven *et al.* (1991), treat the Pell equation as part of the theory of continued fractions, which provides an interesting, but less elementary, approach.

Example 4.1.4 Consider the hyperbola $H : 2x^2 - 7xy + 4y^2 = 4$.

For the rational points, start from the integer point $(0, 1)$ and find where the lines $x = 0$ and $y = 1 + mx$, where m is rational, meet the hyperbola again. Putting $m = u/v$, where u and v are coprime integers, we find the rational points to be $(0, 1)$, $(0, -1)$ from the line $x = 0$, and

$$(x, y) = \left(\frac{7v^2 - 8uv}{2v^2 - 7uv + 4u^2}, \frac{2(v^2 - 2u^2)}{2v^2 - 7uv + 4u^2} \right), \tag{4.1.6}$$

from the line $y = 1 + mx$.

To find the integer points is a much more complicated task. The first thing to do is to transform the equation, by completing the square, into the more manageable form $w^2 - 17y^2 = 32$, where $w = |4x - 7y|$. At this point, one might hope that a theorem such as Theorem 4.1.3 would enable us to find all positive integer solutions by means of an iteration of some basic solution. However, when the right-hand side of the equation is not 1, although sequences of solutions may exist, there is often more than one such sequence. There are, in this case, four basic solutions, basic in the sense that they cannot be obtained from each other by iteration. These are $y = 1$, $w = 7$, and $x = 0$; $y = 2$, $w = 10$, and $x = 6$ or 1; $y = 14$, $w = 58$, and $x = 10$ or 39; and $y = 23$, $w = 95$, and $x = 64$. Some of these integer points are shown in Fig. 4.3.

In order to generate the four sequences of solutions from these basic solutions we need to have the smallest solution of the auxiliary equation $w^2 - 17y^2 = 1$. This is $w = 33$ and $y = 8$. We then form the matrix $\boldsymbol{A}$ with first row $(33, 8)$ and second row $(136, 33)$. Here the entry 136 is chosen so that $\det \boldsymbol{A} = 1$. Then, if (w_0, y_0) is one of the basic solutions, then the sequence that derives from it is found from the equations $(w_m, y_m) = (w_0, y_0)\boldsymbol{A}^m$, $m = 0, 1, 2, \ldots$. The four solutions are $(w_1, y_1) = (367, 89)$, $(602, 146)$, $(3818, 926)$, and $(6263, 1519)$. These provide the integer points $(x, y) = (64, 89)$, $(105, 146)$, $(406, 146)$, $(2575, 926)$, $(666, 926)$, and $(4224, 1519)$ on H. The reader may check that these are indeed solutions, but no proof is given that all solutions are obtained by this method.

Example 4.1.5 (Introducing singular points) Consider the conic $C : f(x, y) = y^2 - 2x^2 + xy + 7x - 4y - 5 = 0$.

This example gives an opportunity to introduce the concept of a *singular point*. Pictorially, a singular point is where a curve crosses over itself. For curves $f(x, y) = 0$ of the second degree, the only possibility is when the conic degenerates into a pair of straight lines, their point of intersection being the singular point.

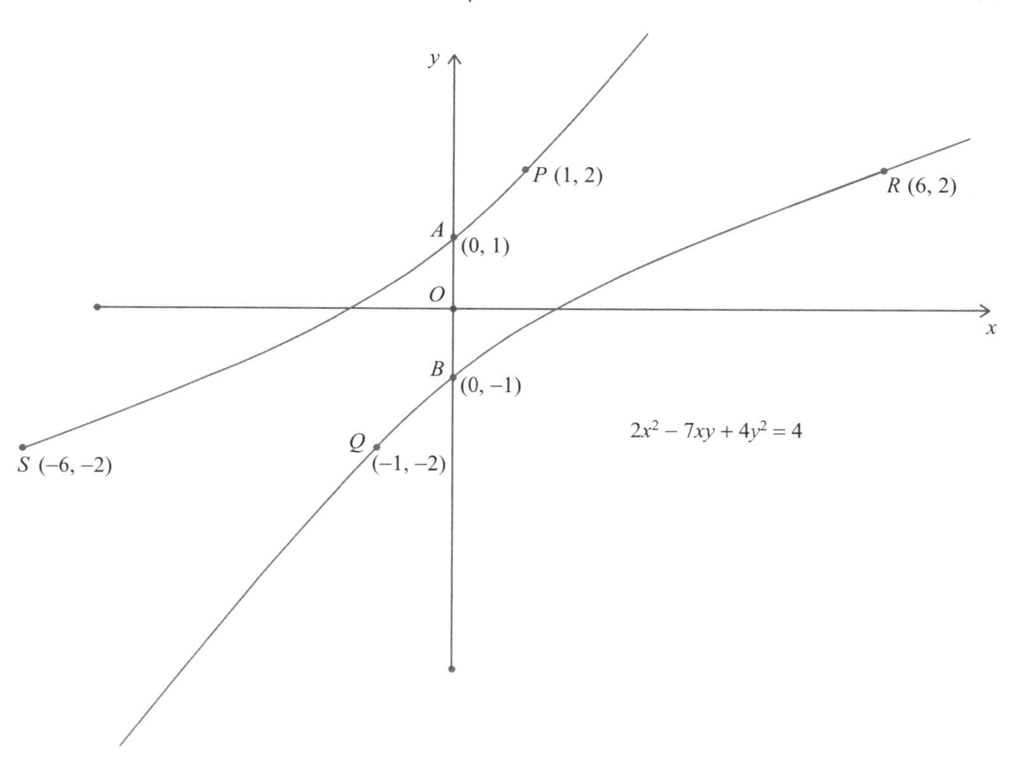

Fig. 4.3 Integer points on a hyperbola.

Here $f(x, y) = (y - x + 1)(y + 2x - 5)$ and the point where both of these terms vanish is $(x, y) = (2, 1)$. We see this clearly if we write

$$\begin{aligned} f(x, y) &= [(y-1) - (x-2)][(y-1) + 2(x-2)] \\ &= (y-1)^2 + (x-2)(y-1) - 2(x-2)^2 . \end{aligned} \tag{4.1.7}$$

In other words, if we express $f(x, y)$ as a two-variable Taylor series about a singular point then the absolute term and the terms of first degree in that expansion vanish. This observation gives the clue as to how to find a singular point, and provides an analytic definition. This is to work out the partial derivatives $\partial f/\partial x$ and $\partial f/\partial y$, and solve the simultaneous equations $\partial f/\partial x = 0$ and $\partial f/\partial y = 0$. Any points (x, y) which emerge that also satisfy $f(x, y) = 0$ are the singular points. Points that are not singular are called *simple*. If one has a singular point but the second partial derivatives do not vanish then the singular point is called a *double point*. If a curve contains no singular points (even at infinity when the curve is embedded in the projective plane) then the curve is termed *non-singular*. In higher-order curves a *triple point* may be defined where the second-order derivatives vanish, and so on.

Exercises 4.1

4.1.1 Starting from the point $(8, 1)$, find all rational points on the curve with equation $x^2 + y^2 = 65$. Also find all of the integer points.

4.1.2 Find expressions for all rational points and all integer points on the hyperbola with equation $x^2 - 7y^2 = 1$. Are there any integer points on the hyperbola $7x^2 - y^2 = 1$?

4.1.3 Find the least positive integer solution of the equation $x^2 - 18y^2 = 1$. Is there any integer solution of the equation $x^2 - 18y^2 = -1$?

4.1.4 Find any integer points on the hyperbola with equation $x^2 - 4y^2 = 13$.

4.1.5 Show that, for certain values of m and c, the line $y = mx + c$ meets the conic with equation $y^2 - 4x^2 + 6x = 9$ in more than two points.

4.2 Rational points on cubic curves with a singular point

The projective plane

Before proceeding with examples, we first show how to embed curves defined in $\mathbb{R}^2$ so that they become defined in the real projective plane $\mathbb{P}^2$. This is necessary if we are to have a proper classification of cubic curves, and in particular of elliptic curves.

If $f(x, y)$ is a polynomial of degree three, then we define the function $F(X, Y, Z) = f(X/Z, Y/Z)Z^3$. In this way, the cubic curve $f(x, y) = 0$ is transformed into a homogeneous cubic $F(X, Y, Z) = 0$ in $\mathbb{R}^3$. For example, $x^3 + y^3 = 7$ becomes $X^3 + Y^3 = 7Z^3$. We now define an equivalence class on the points of $\mathbb{R}^3\backslash\{(0, 0, 0)\}$ by saying that (X, Y, Z) and (kX, kY, kZ) are equivalent for all nonzero real numbers k. The reason that this is fruitful is because of the homogeneous nature of F, whereby if (X, Y, Z) lies on $F(X, Y, Z) = 0$ then so does (kX, kY, kZ). We now identify the lines through the origin in which $X : Y : Z$ are in fixed ratio as points of the projective plane $\mathbb{P}^2$. Points (x, y) on $f(x, y) = 0$ may be identified on $F(X, Y, Z) = 0$ by $X : Y : Z = x : y : 1$, but there are now points at infinity on $f(x, y) = 0$, which may be identified by lines on which $Z = 0$. For example, $1 : -1 : 0$ lies on $X^3 + Y^3 = 7Z^3$, but does not correspond to any finite point of the curve in the original real plane.

In the remainder of this section we consider curves that possess a singularity either at a finite point or at infinity. In Example 4.2.1 there is a double point at infinity, in Example 4.2.2 there are two double points, and in Example 4.2.3 there is one double point. Methods of finding rational and integer points differ depending on the nature of the singularities.

Example 4.2.1 Consider the curve $C : y = x^3$.

Trivial as this example is, it serves to illustrate the use of singular points in obtaining parameters for rational points on a cubic curve. There are no finite singular points, but in the projective plane the curve becomes $YZ^2 = X^3$. For singular points we have $3X^2 = 0$, $Z^2 = 0$, and $2YZ = 0$, providing a double point $X : Y : Z = 0 : 1 : 0$. Lines through the double point are of the form $X = cZ$, for real c. Back in the real plane these correspond to the lines $x = c$ parallel to the y-axis. When $x = c$, we have $y = c^3$, and this parameterisation provides rational (integer) points on the curve for rational (integer) c.

Example 4.2.2 Consider the curve $C : f(x,y) = y^3 - 2x^3 + yx^2 - 2xy^2 - 5y + 10x = 0$.

For singular points,

$$3y^2 + x^2 - 4xy - 5 = 0 \tag{4.2.1}$$

and

$$-6x^2 + 2xy - 2y^2 + 10 = 0\,. \tag{4.2.2}$$

Eliminating x^2 we find that $x = (8y^2 - 10)/11y$. Substituting back and solving we obtain $y = (1/3)\sqrt{3}$ and $x = (2/3)\sqrt{3}$, $y = -(1/3)\sqrt{3}$ and $x = (2/3)\sqrt{3}$, $y = 2$ and $x = 1$, and $y = -2$ and $x = -1$. Of these four points, only $(1,2)$ and $(-1,-2)$ lie on the curve, and hence we have two finite double points. It seems clear that the line joining them, which has equation $y = 2x$, since it encompasses four intersections with a curve of degree three, must be part of the curve itself, which is therefore the union of a line and a conic. In fact, $f(x,y) = (y - 2x)(x^2 + y^2 - 5)$. The integer points on the curve are therefore $x = c$ and $y = 2c$, for any integer c, and $(2,1)$, $(2,-1)$, $(-2,1)$, $(-2,-1)$, $(1,2)$, $(1,-2)$, $(-1,2)$, and $(-1,-2)$. For the rational points, $x = c$ and $y = 2c$, for c rational, together with

$$x = \frac{2(u^2 + uv - v^2)}{u^2 + v^2}\,, \tag{4.2.3}$$

$$y = \frac{u^2 - 4uv - v^2}{u^2 + v^2}\,, \tag{4.2.4}$$

where u and v are coprime integers, and also those points with one or both of the co-ordinates being the negative of these. The methods of Section 4.1 were used to obtain the rational points on the circle $x^2 + y^2 = 5$, see Fig. 4.4.

Example 4.2.3 Consider the curve $C : y^2 = x^3 - x^2 - 5x - 3$.

For singular points, $y = 0$ and $3x^2 - 2x - 5 = 0$. The two possibilities are $(5/3, 0)$ and $(-1, 0)$. Of these, only the second one lies on the curve, and hence we have one double point. The method of obtaining a parameterisation is now to consider lines through the double point with rational slope, and then their intersections with the

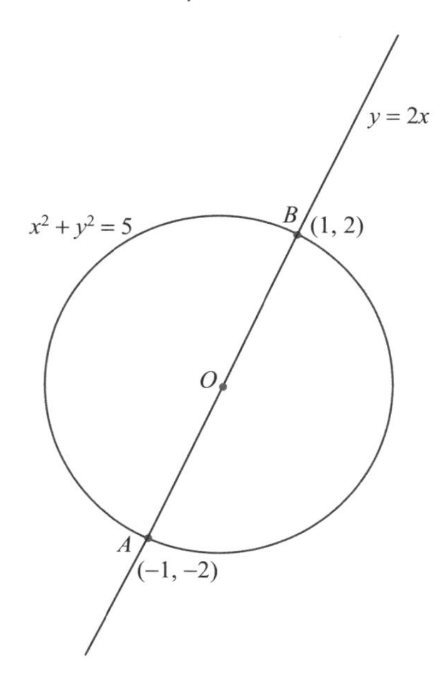

Fig. 4.4 A cubic with two double points.

curve give all of the rational points on it. These lines are of the form $y = m(x+1)$, where m is rational. They meet the curve again at the points with co-ordinates

$$\begin{aligned} x &= m^2 + 3\,, \\ y &= m(m^2 + 4)\,. \end{aligned} \tag{4.2.5}$$

These equations, by varying m, together with $x = -1$ and $y = 0$, give the co-ordinates of all of the rational and integer points.

Exercises 4.2

4.2.1 Find all integer points lying on the curve with equation $x^3 + y^3 = (x+y)^2$.

4.2.2 Prove that, if the curve with equation $y^2 = p(x)$, where $p(x)$ is a cubic polynomial with integer coefficients, has a double point, then the equation $p(x) = 0$ has a double root, and that this root has a rational value.

4.2.3 Find the integer points on the curve with equation $y^2 = x^3 + 3x^2$.

4.3 Elliptic curves

An *elliptic curve* over the field of real numbers is a curve of degree three with equation $f(x,y) = 0$ such that $f(x,y)$ is irreducible over the real numbers and has no singular point in the real projective plane $\mathbb{P}^2$. Elliptic curves may be defined over other fields in the same way.

If a cubic curve has an equation that factors over the real field, then one of the factors must be linear, and so the curve contains a line, see Example 4.2.2. Furthermore, if a cubic curve has a singularity then it is impossible to determine a unique tangent at that point. Consider, for example, the curve with equation $x^3 + y^3 = 3axy$, where a is a real constant. This curve has a singularity at $(0, 0)$, where the curve crosses itself. Both branches of the curve are smooth at the origin, but there is no unique tangent there. Elliptic curves are defined to exclude both of these eventualities.

The definition of an elliptic curve imposes the requirement that it should have the following property. If you join two distinct points of an elliptic curve or draw the tangent at a given point, then the line must intersect the curve at a unique well-defined point. Thus, if P and Q are two points on the curve, then the line through P and Q must meet the curve again at a *unique* point R and we may write $R = P * Q = Q * P$. This property cannot hold if the curve contains a line, which is why the requirement is made on $f(x, y)$ that it should be irreducible. Nor can the property hold if there is a singular point P, because we cannot then decide what is meant by $P * P$. On a non-singular curve, $P * P$ is defined as the *unique* point on the curve other than P lying on the tangent at P. An elliptic curve is embedded in the projective plane because we require the same properties to hold whether the points involved are at infinity or not.

It is, of course, possible that $P * Q = P$, when PQ is tangent at P; it is also possible that $P * Q = Q$, when PQ is tangent at Q.

Before embarking on the algebra of the $*$ operation, we give two numerical examples, the first of which is a classic exercise. These examples also provide numerical material to illustrate the general theory.

Example 4.3.1 Consider the curve $C : x^3 + y^3 = 7$.

In the projective plane this becomes $X^3 + Y^3 = 7Z^3$. For singular points we have $X = Y = Z = 0$, and as this point does not belong to $\mathbb{P}^2$ we deduce that there are no singular points. There is a point at infinity on the curve, which we denote by J; its co-ordinates are $1 : -1 : 0$. There are also two obvious integer points $P(2, -1)$ and $Q(-1, 2)$. In $\mathbb{P}^2$ the line PQ has equation $X + Y = Z$, and this meets the curve again at J and at no other point, since $X^2 - XY + Y^2 = 7(X + Y)^2$ merely reproduces P and Q.

We now show how to find an infinite sequence of rational points on the curve. This is done by starting at $P_0 = P$ and finding the point P_1 where the tangent at P_0 meets the curve again. We then do the same for P_1 to find a new rational point P_2, and so on. Finally, we show that P_k, $k = 0, 1, 2, \ldots$, are distinct. We proceed by induction. Straightforward calculus shows that the equation of the tangent at $P_n(x_n, y_n)$ has equation

$$y_n^2 y + x_n^2 x = 7\,. \tag{4.3.1}$$

Since $x_n^3 + y_n^3 = 7$, this meets the curve again where

$$(7 - x_n^2 x)^3 = (7 - x^3)(7 - x_n^3)^2\,, \tag{4.3.2}$$

that is, where $x = x_n$ (a double root) and

$$x = x_{n+1} = \frac{(14 - x_n^3)x_n}{2x_n^3 - 7} = \frac{x_n(2y_n^3 + x_n^3)}{x_n^3 - y_n^3}, \tag{4.3.3}$$

with a similar expression for y_{n+1} with x_n and y_n interchanged. In the projective plane this provides the recurrence relations

$$\begin{aligned} X_{n+1} &= X_n(2Y_n^3 + X_n^3), \\ Y_{n+1} &= -Y_n(2X_n^3 + Y_n^3), \\ Z_{n+1} &= Z_n(X_n^3 - Y_n^3). \end{aligned} \tag{4.3.4}$$

The start of the induction is P_0 with co-ordinates $2 : -1 : 1$, which certainly lies on the curve. Hence all of the points P_k lie on the curve. Since X_0 is even and Y_0 and Z_0 are odd, it follows by induction that X_n is even and Y_n and Z_n are odd. Evidently, $X_n | X_{n+1}$ and, since $2Y_n^3 + X_n^3 = 2 \pmod 4$, X_{n+1} has one more power of 2 than X_n. It follows that the P_k, $k = 0, 1, 2, \ldots$, are distinct. We give the co-ordinates of P_1, P_2, and P_3. These are $4 : 5 : 3$, $1256 : -1265 : -183$, and $-2\,596\,383\,146\,704 : 2\,452\,184\,545\,855 : -733\,037\,580\,903$. Figure 4.5 illustrates the construction of P_1 from P. A similar sequence arises if we start with Q rather than P. It is worth noting that there are other rational points on the curve. For example, if we join Q to P_1 then we have the line with equation $7Y + X = 13Z$, and this meets the curve again at $R_{1,0}$, whose co-ordinates are $-17 : 73 : 38$. Note also that the tangent at the point J in the projective plane (the asymptote in the real plane) has triple point contact. Points such as these are called *inflection points*. Note that the line joining $P_1(4 : 5 : 3)$ and $Q_1(5 : 4 : 3)$ has equation $X + Y = 3Z$ and meets the curve again at J, so with an obvious notation $R_{1,1} = J$. In terms of the $*$ operation, we have $P_k * Q_k = J$, $J * J = J$, $P_k * P_k = P_{k+1}$, $Q_k * Q_k = Q_{k+1}$, and $P_k * Q_l (k \neq l) = R_{k,l}$.

Example 4.3.2 Consider the curve $C : y^2 = x^3 + 4x + 9$.

It is easily checked that this is an elliptic curve. There are six easily recognised integer points, namely $A(-1, 2)$, $B(0, 3)$, $C(2, 5)$, $D(-1, -2)$, $E(0, -3)$, and $F(2, -5)$. In the projective plane the point $J(0 : 1 : 0)$ is an inflection point. Lines parallel to the y-axis all pass through J. So, for example, $A * D = J$, $B * E = J$, and $C * F = J$. The line BF has equation $y + 4x = 3$ and meets the curve again at $G(14, -53)$. This means that $H(14, 53)$ also lies on the curve, and we have $B * F = G$. The line BD has equation $y = 5x + 3$ and this meets the curve again at $K(26, 133)$. This means that $L(26, -133)$ also lies on the curve, and we have $B * D = K$.

The $*$ and $+$ operations on an elliptic curve

We have already introduced the $*$ operation, which on an elliptic curve is a closed binary operation. It is also commutative since it is obvious that PQ and QP, being the

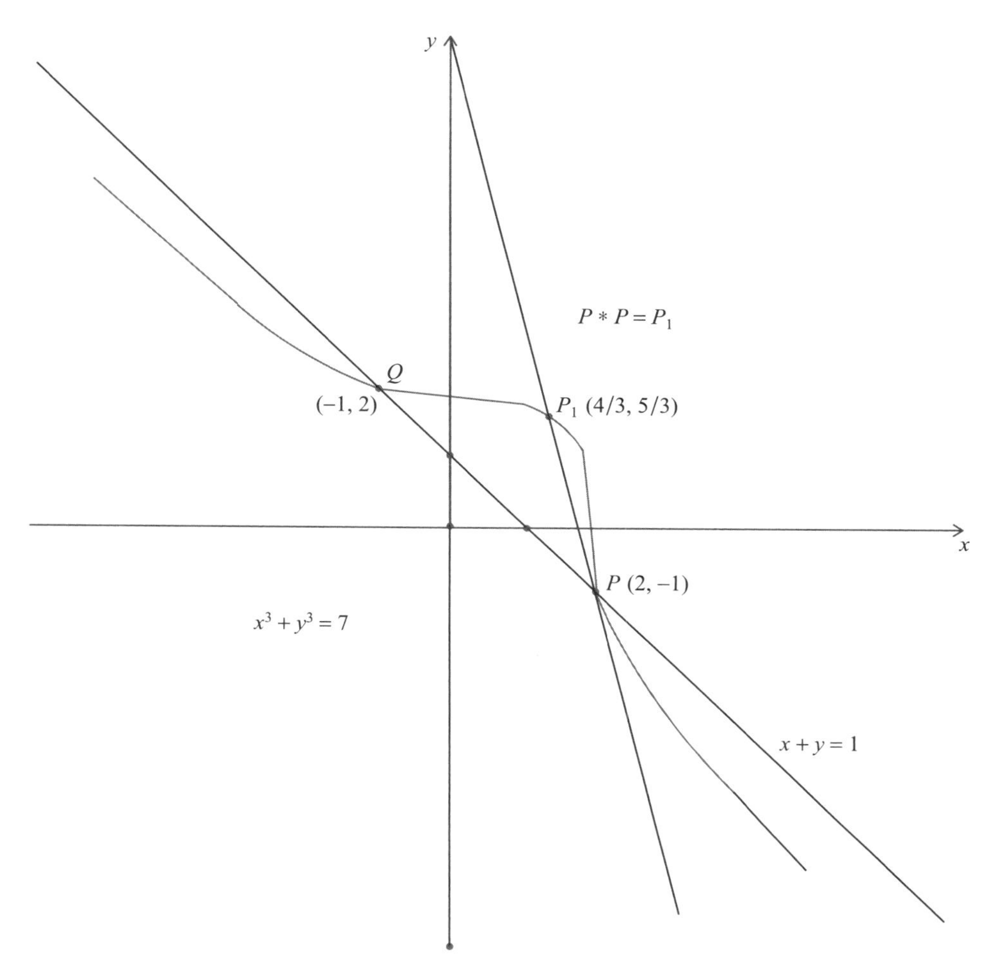

Fig. 4.5 An elliptic curve showing $P * P = P_1$.

same line, cut the curve again at one and the same third point, and hence $P*Q = Q*P$. It is also the case that $P * (P * Q) = Q$ since, if PQ meets the curve at R, then PR meets the curve at Q. However, there is no point E on the curve such that $P * E = P$ for all P on the curve. Indeed, $P * E = P$ if and only if PE is tangent at P. That is, if and only if $E = P * P$, and this point is dependent on P. Also, Q is a point of inflection if and only if $Q * Q = Q$.

It is possible, however, to define a binary operation $+$ that is closed, associative, commutative, and which possesses both an identity and inverses. Then, with respect to $+$, the points on the elliptic curve form an Abelian group. In order to define $+$, choose *any* point E on the curve and, given any two points P and Q on the curve, define $P+Q = E*(P*Q)$. Closure is obvious. Also, $+$ is commutative since $P*Q = Q*P$. E is an identity since $P + E = E + P = E * (E * P) = P$. The inverse of P is

$P*(E*E)$ since $P+P*(E*E) = E*(P*(P*(E*E))) = E*(E*E) = E$; the inverse is unique since, if $P+Q = E$, then $E*(P*Q) = E$ and hence $P*Q = E*E$, that is, $Q = P*(E*E)$.

The symbol $-$ is used in an obvious way. Thus $P - Q = R$ means the same as $P = R + Q$. Multiplication by an integer also has an obvious meaning. For example, $2P$ means $P + P$. Evidently, by changing the choice of E, the binary operation $+$ is also changed. However, see Theorem 4.3.4 below, the Abelian groups obtained with different choices of E are isomorphic to one another. Without this property the concept of the group would not be important.

As an example of the above, let us return to Example 4.3.2 with $E(0, -3)$. The tangent at E has equation $3y + 2x + 9 = 0$ and this meets the curve again at $(4/9, -89/27)$, and this is the point $E*E$. Now $B*G = F$ and hence $B+G = E*(B*G) = E*F$. Now the equation of the line joining E and F is $y+x+3 = 0$ and this meets the curve again at $D(-1, -2)$. Hence $B + G = D$. We now work out the inverse of B. According to the above, this is the point $B*(E*E)$. The equation of the line joining $(0, 3)$ and $(4/9, -89/27)$ is $6y + 85x = 18$ and this meets the curve again at $B_0(801/4, -22\,671/8)$, the inverse of B. We next find that $B_0 * D = (-61/49, 496/343)$. The line joining E and $B_0 * D$ has equation $7y + 25x + 21 = 0$ and this meets the curve again at $(14, -53)$, which is the point G. We have shown that

$$B_0 + (B + G) = B_0 + D = G = (B_0 + B) + G\,. \tag{4.3.5}$$

This illustrates the associative law.

The associative law

Theorem 4.3.3 *Let $C(f) : f(x, y) = 0$ be an elliptic curve over the real field, E be an arbitrary origin on $C(f)$, and P, Q, and R be any three points on $C(f)$. Then, with the operation $+$ defined by $P + Q = E*(P*Q)$, the associative law $P + (Q + R) = (P + Q) + R$ holds.* □

For the proof we refer to Niven *et al.* (1991).

The group G defined for $C(f)$ with respect to an origin E is isomorphic to the group H defined for $C(f)$ with respect to a different origin F.

Theorem 4.3.4 *Let E and F denote two points on $C(f)$. For A and B in $C(f)$, let $A + B$ denote the addition with E as origin and $A \wedge B$ denote the addition using F as origin. Then $A \wedge B = A + B - F$ and, furthermore, the group $G(f) = \{C(f), +\}$ is isomorphic to the group $H(f) = \{C(f), \wedge\}$.*

Proof We have $A \wedge B + F = F*(A*B) + F = E*(F*(F*(A*B))) = E*(A*B) = A + B$. The mapping t from $G(f)$ to $H(f)$ defined by $t(A) = A + F$ satisfies $t(A+B) = A+B+F = (A+F)+(B+F)-F = (A+F)\wedge(B+F) = t(A)\wedge t(B)$, so t is a homomorphism. However, t is clearly 1–1 and onto, and is hence an isomorphism. □

If $C(f)$ has any rational points, then E may be chosen to be such a point, and, if we restrict attention only to rational points of $C(f)$, then the group involved is a subgroup of $G(f)$.

It may be shown, by a famous theorem of Mordell, that the group of rational points on an elliptic curve is a finitely-generated Abelian group.

Exercises 4.3

4.3.1 In Example 4.3.2 show that $E * D = F$ and $B * C = A$.

4.3.2 In Example 4.3.2 show that $B_0 * D = (-61/49, 496/343)$.

4.3.3 In Example 4.3.1 take the point at infinity to be the origin E (labelled as J previously), and show that $P_1 = -2P_0 = -Q_1$, and more generally that $P_k = (-2)^k P_0$. Find $3P_0$ and show how to obtain nP_0.

4.3.4 Consider the elliptic curve with equation $2x(x^2-4) = y(y^2-1)$. Let $(0,0)$ be the origin E. Define the points $A(0,1)$, $B(2,0)$, and $C(2,1)$. Show that $A+B$ has co-ordinates $(28/17, -31/17)$ and that $B+C$ has co-ordinates $(-2,1)$. Find the common co-ordinates of $(A+B)+C$ and $A+(B+C)$.

4.3.5 If A, B, and C are three distinct points of $C(f)$ then prove that A, B, and C are collinear if and only if $A+B+C = E * E$.

4.3.6 If A and B are two distinct points of $C(f)$ and AB is tangent at B then prove that $A + 2B = E * E$.

4.3.7 Find all integer points on the elliptic curve $x^3 + y^3 = 1$, and explain why the chord and tangent method yields no further rational points.

4.3.8 For what values of k is the curve $x(x^2-4) = ky(y^2-1)$ an elliptic curve?

4.4 Elliptic curves of the form $y^2 = x^3 - ax - b$

In this section we consider curves of the form $y^2 = x^3 - ax - b$, where a and b are integers. The importance of curves of this form lies in the fact that, if $C(f)$ is an elliptic curve with a rational point and $f(u,v)$ has rational coefficients, then it may be transformed by what is called a birational transformation (the variables u and v being the quotient of two polynomials in x and y with rational coefficients, and the transformation having an inverse) into an equation of the above form. This form is called the *Weierstrass normal form.*

In the projective plane this curve has equation

$$F(X,Y,Z) = X^3 - aXZ^2 - bZ^3 - Y^2Z = 0\,. \tag{4.4.1}$$

For singular points the partial derivatives vanish, that is,

$$
\begin{aligned}
3X^2 - aZ^2 &= 0\,, \\
-2YZ &= 0\,, \\
2aXZ + 3bZ^2 + Y^2 &= 0\,.
\end{aligned}
\qquad (4.4.2)
$$

If $Z = 0$ then $X = Y = 0$, and this is not a point of the projective plane. If $Y = 0$ then $X = -3bZ/2a$ and $3X^2 = aZ^2$, so that $4a^3 - 27b^2 = 0$. Those familiar with cubic equations will recognise $D = 4a^3 - 27b^2$ as the discriminant, and its vanishing is the condition for the equation $p(x) = x^3 - ax - b = 0$ to have a repeated root. Further, if $D > 0$ then the equation $p(x) = 0$ has three real roots, and if $D < 0$ then $p(x) = 0$ has only one real root. It follows that if $D > 0$ then we have an elliptic curve with two connected components, one a closed oval and the other extending to the point $(0 : 1 : 0)$ at infinity. If $D = 0$ then we do not have an elliptic curve, and if $D < 0$ then we have an elliptic curve with one connected component. Connectedness depends on the field one is working with, and we are supposing here that our curves are defined over the real field.

We now consider a number of special cases.

Example 4.4.1 Let $C_1 : y^2 = x^3 + 4x$ and $C_2 : y^2 = x^3 - x$.

We prove that on C_1 the only finite points with rational co-ordinates are $A(0,0)$, $B(2,4)$, and $C(2,-4)$, and that on C_2 the only finite points with rational co-ordinates are $P(0,0)$, $Q(1,0)$, and $R(-1,0)$. In both cases there is the point at infinity $E(0 : 1 : 0)$. On C_1 we have the group structure on rational points defined by $4B = 4C = 2A = E$, $A + B = C$, and $A + C = B$, the cyclic group of order four. On C_2 we have the group structure on rational points defined by $P + Q + R = E$ and $2P = 2Q = 2R = E$, the direct sum of two copies of the cyclic group of order two.

In order to show that the above points are the only rational points, we note that $(0,0)$ lies on each curve. Then, since $x = 0$ is the tangent line in each case, if there is any other rational point on either curve then it must lie on a line with equation $y = mx$, for some rational value of m. On C_1 the line $y = mx$ has intersection points $x = 0$ or $x^2 - m^2x + 4 = 0$, and this leads to rational points if and only if $m^4 - 16$ is the square of a rational; that is, if and only if $m = 2$ or $m = -2$ (leading to the rational points B and C), or nonzero integers x, y, and z exist such that

$$x^2 + 16y^4 = z^4\,. \qquad (4.4.3)$$

On C_2 the line $y = mx$ has intersection points $x = 0$ (leading to the rational points Q and R) or $x^2 - m^2x - 1 = 0$, and this leads to rational points if and only if $m^4 + 4$ is the square of a rational; that is, if and only if $m = 0$ or nonzero integers u, v, and w exist such that

$$u^4 + 4v^4 = w^2\,. \qquad (4.4.4)$$

The curves C_1 and C_2, together with their rational points, are shown in Figs 4.6 and 4.7, respectively.

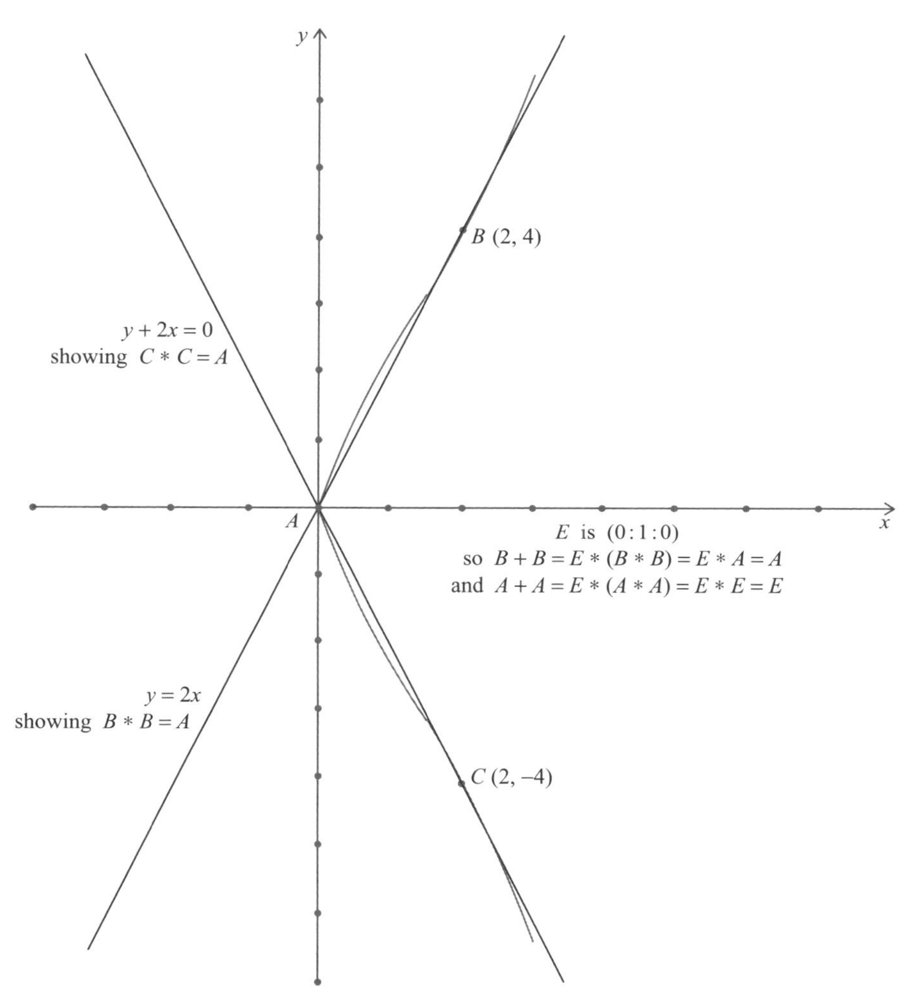

Fig. 4.6 The elliptic curve with equation $y^2 = x^3 + 4x$ and with one connected component.

Method of infinite descent

To complete Example 4.4.1 we prove that neither of the Diophantine equations $u^4 + 4v^4 = w^2$ nor $x^2 + 16y^4 = z^4$ have solutions in which all of u, v, w, x, y, and z are positive integers. It is a good example of an argument called the method of *infinite descent*. The method supposes that such solutions exist and, if they do, then there must be a solution in each case in which w and z have least positive values. We then show that any solution leads to another solution in which u, v, w, x, y, and z are positive, but in which w and z are smaller than before. This leads to a contradiction, establishing the non-existence of such solutions.

Following this plan, suppose that u, v, and w are positive integers satisfying

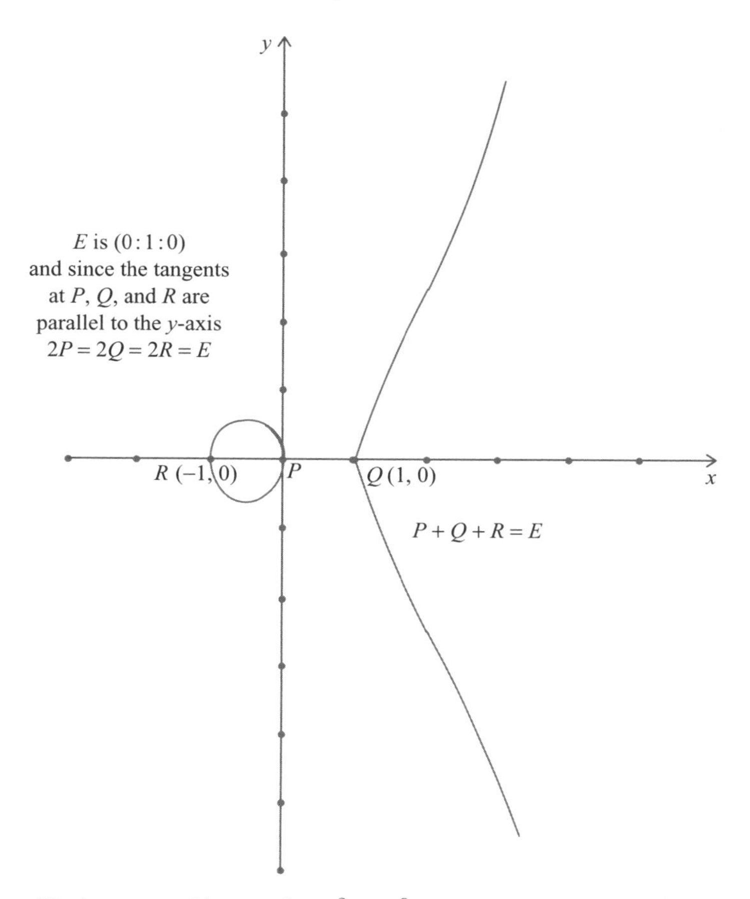

Fig. 4.7 The elliptic curve with equation $y^2 = x^3 - x$ and with two connected components.

$u^4 + 4v^4 = w^2$, where we may suppose that any common factor has been removed. Then $(u^2, 2v^2, w)$ forms a primitive Pythagorean triple, and so from Theorem 1.1.2 coprime integers r and s of opposite parity exist such that

$$u^2 = r^2 - s^2, \quad 2v^2 = 2rs, \quad w = r^2 + s^2. \tag{4.4.5}$$

The first of these shows that (u, s, r) is also a primitive Pythagorean triple, so r is odd and s is even. Since $rs = v^2$, then, as r and s are coprime, positive integers y and z exist such that

$$r = z^2, \quad s = 4y^2, \quad v = 2yz. \tag{4.4.6}$$

Writing $u = x$ the equation $u^2 + s^2 = r^2$ leads to positive integers x, y, and z satisfying $x^2 + 16y^4 = z^4$. Furthermore, $z \leqslant z^4 = r^2 < r^2 + s^2 = w$, so $z < w$. Now suppose that x, y, and z are positive integers satisfying eqn (4.4.3), where we may suppose that any common factor has been removed. Then $(x, 4y^2, z^2)$ forms a primitive Pythagorean triple, and so from Theorem 1.1.2 coprime integers h and k of opposite parity exist such that

$$x = h^2 - k^2\,, \quad 4y^2 = 2hk\,, \quad z^2 = h^2 + k^2\,. \tag{4.4.7}$$

It does not matter which of h or k is even, so, since $hk = 2y^2$, we suppose without loss of generality that $h = u'^2$ and $k = 2v'^2$; then, setting $z = w'$, we have $u'^4 + 4v'^4 = w'^2$. Moreover, $w' = z < w$. Continuing this process, we find another positive solution x', y', and z' with $0 < z' < w' \leqslant z < w$. This infinite descent leads to a contradiction to the fact that there must be a solution with least positive w and z.

Example 4.4.2 Consider the curve $C : y^2 = x^3 - 9$.

We prove that there are no integer points on the curve C. If x is even then the right-hand side is 3 (mod 4) and so cannot be a perfect square. So, if there is a solution, then x must be odd and y even. From this we see that $x = 1$ (mod 4). Now, $y^2 + 1 = x^3 - 8 = (x-2)(x^2 + 2x + 4)$. However, if y is even then every factor of $y^2 + 1$ must be 1 (mod 4). But $x - 2 = 3$ (mod 4). This contradiction establishes the result.

Exercises 4.4

4.4.1 Find all rational points on the curve with equation $y^2 = x^3 + x$.

4.4.2 Find all rational points on the curve with equation $y^2 = x^3 - 4x$.

4.4.3 Prove that there are no integer points on the curve with equation $y^2 = x^3 + 23$.

4.4.4 Let $P(r, s)$ be a point on the elliptic curve with equation $y^2 = x^3 - ax - b$, where a and b are integers. If r and s are integers, then find the condition that $2P$ should have integer co-ordinates.

5 Shapes and numbers

The 'numbers' in the title of this chapter are actually integers, but the title is determined by the fact that well-known names for what we consider here are such things as triangular numbers, polygonal numbers, and polyhedral numbers. In this chapter we try to provide material that is either new or less well known, rather than provide what tends to be given in books on recreational mathematics.

We start by considering the triangular numbers in some detail, highlighting their connection with the squares, and in particular dealing with the representation of positive integers as the sum of triangular numbers. In doing this we draw attention to the connections that exist between theorems on triangular numbers and square numbers. We point out that any primitive Pythagorean triplet also has its analogue amongst the triangular numbers. We then describe other shapes and the numbers associated with them. Next we consider the problem of determining which polygonal numbers are also square numbers or triangular numbers. For some polygonal numbers only a finite number of them are squares, but for others there are an infinite number. The Pell equation once again features in this problem. Finally, we give a brief account of the Catalan numbers.

5.1 Triangular numbers

The *triangular numbers* are the integers T_n that may be described pictorially by n rows of dots in the shape of a triangle. T_6 is illustrated in Fig. 5.1. The definition of T_n is that it is equal to the number of dots in a triangle made up of n rows of dots, with one dot in the first row, two dots in the second row, and so on, with finally n dots in the nth row. Hence

$$T_n = 1 + 2 + \ldots + n = \frac{1}{2}n(n+1)\,. \tag{5.1.1}$$

We may extend the definition of T_n algebraically to negative integers by noting that $T_{-n-1} = T_n$, since $\frac{1}{2}(-n-1)(-n) = \frac{1}{2}n(n+1)$.

Triangular numbers are closely related to square numbers by virtue of the equation

$$8T_n + 1 = (2n+1)^2\,. \tag{5.1.2}$$

This means that the equation $x^2 = 2N + 1$ is solvable for integral x and N if and only if $N = 4T_n$ for some integer n. From this, we see that the equation $x^2 = M$ is solvable for integral x and M if and only if $M = 2^{2k}(8T_n + 1)$ for some positive integers k and n.

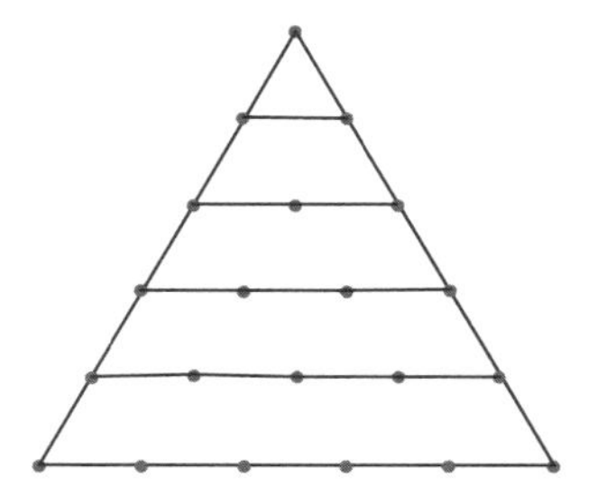

Fig. 5.1 The triangular number $T_6 = 21$.

Theorem 5.1.1 *The equation $T_n = A$ is solvable for integers n and A if and only if there exists a positive integer x such that $x^2 = 8A+1$, and when this is so $n = \frac{1}{2}(x-1)$ or $-\frac{1}{2}(x+1)$.* □

The proof is left to the reader

In other words, the representation of an integer by a triangular number is in correspondence with the representation of a related integer as a perfect square. We now show that the same consideration applies in relation to the problem of which integers can be represented as the sum of two triangular numbers.

Theorem 5.1.2 *The equation $T_l + T_m = B$ is solvable for integers l, m, and B if and only if integers x and y exist such that $x^2 + y^2 = 4B + 1$, and when this is so we may choose $l = \frac{1}{2}(x + y - 1)$ and $m = \frac{1}{2}(x - y - 1)$.*

Proof If

$$T_l + T_m = B\,, \tag{5.1.3}$$

then

$$\begin{aligned} 4B + 1 &= 2l(l + 1) + 2m(m + 1) + 1 \\ &= (l + m + 1)^2 + (l - m)^2 = x^2 + y^2\,, \end{aligned} \tag{5.1.4}$$

where $x = l + m + 1$ and $y = l - m$. Conversely, if $x^2 + y^2 = 4B + 1$, then necessarily one of x and y must be even and the other odd. Then $l = \frac{1}{2}(x + y - 1)$ and $m = \frac{1}{2}(x - y - 1)$ are integers such that $T_l + T_m = B$. □

It should be noted that, for certain values of B in Theorem 5.1.2, there may be several sets of values of l and m that can be found to satisfy eqn (5.1.3). For example, when $B = 16$ we could have $(x, y) = (8, 1)$ or $(7, 4)$, with corresponding values of $(l, m) = (4, 3)$ or $(5, 1)$, respectively.

Note that, with T_n defined for negative integers, as at the start of this section, Theorem 5.1.2 is true whether the integers l and m appearing are positive, zero, or negative. For example, when $B = 16$, if we take $(x, y) = (1, 8)$ then we find that $(l, m) = (4, -4)$.

We now introduce a binary operation $*$ on the integers and prove that the operation is closed on integers that can be expressed as the sum of two triangular numbers.

Theorem 5.1.3 *Define the binary operation $*$ on the integers by the rule*

$$B * C = 4BC + B + C\,. \tag{5.1.5}$$

Then

$$(T_l + T_m) * (T_p + T_q) = T_L + T_M\,, \tag{5.1.6}$$

where

$$L = lp + mq - lq + mp + p + m \quad \textit{and} \quad M = lp + mq + lq - mp + l + q\,. \tag{5.1.7}$$

Proof Suppose that $T_l + T_m = B$ and $T_p + T_q = C$. Now define $x = l + m + 1$, $y = l - m$, $u = p + q + 1$, and $v = p - q$. Then, by Theorem 5.1.2, we have $x^2 + y^2 = 4B + 1$ and $u^2 + v^2 = 4C + 1$. Then $(xu + yv)^2 + (xv - yu)^2 = (x^2 + y^2)(u^2 + v^2) = (4B + 1)(4C + 1) = 4E + 1$, where $E = B * C$. It follows from Theorem 5.1.2 that $E = T_L + T_M$, where $L = \frac{1}{2}(xu + yv + xv - yu - 1)$ and $M = \frac{1}{2}(xu + yv - xv + yu - 1)$, which reduce to the expressions (5.1.7) when x and y are eliminated in favour of l and m. □

Numbers that may be represented as the sum of two triangular numbers

A theorem that is worth stating, but which we do not prove, is Theorem 5.1.4 below, which is concerned with finding those integers that may be represented as the sum of two triangular numbers. Readers familiar with the problem of finding those integers that may be represented as the sum of two squares will appreciate the connection between the two problems; it is a consequence of Theorem 5.1.2 and the closure properties made explicit in Theorem 5.1.3.

Theorem 5.1.4 *The equation $T_l + T_m = B$ has solutions for l and m if and only if the prime factorisation of $4B + 1$ is of the form $p_1 p_2 \cdots p_t q_1^2 q_2^2 \cdots q_s^2$, where each of the p_j is a prime of the form $4k + 1$ and each of the q_j is a prime of the form $4k + 3$. (The p_js are not necessarily distinct, nor are the q_js.)* □

Pythagoras' theorem using triangular numbers

Suppose that N is any number of the form $4k + 1$ that is representable as the sum of two squares. Then we know from Theorem 5.1.2 that there exist integers l and m such that $4(T_l + T_m) + 1 = N$, and hence that $T_l + T_m = k$. Then

$$(T_l + T_m) * (T_l + T_m) = 4k^2 + 2k = 2T_{2k}\,, \tag{5.1.8}$$

where $k = \frac{1}{2}l(l + 1) + \frac{1}{2}m(m + 1)$.

Moreover, $8T_{2k} + 1 = 16k^2 + 8k + 1 = N^2$. However, the left-hand side of eqn (5.1.8) is also equal to $(T_m + T_l) * (T_l + T_m) = T_L + T_M$, where, by Theorem 5.1.3, $L = l^2 - m^2 + 2lm + 2l$ and $M = m^2 - l^2 + 2lm + 2m$.

We now have the equivalent relations

$$(L + M + 1)^2 + (L - M)^2 = N^2 \tag{5.1.9}$$

and

$$T_L + T_M = 2T_{2k}\,. \tag{5.1.10}$$

It is clear, since $L+M+1$ and $L-M$ are of opposite parity, that, for every Pythagorean triple with one short leg even and the other short leg odd, there exists a relationship between triangular numbers in the form of eqn (5.1.10).

As an example, take $N = 585$, so that $k = 146$. Since $55 + 91 = 146$, we can take $l = 10$ and $m = 13$, leading to $L = 211$ and $M = 355$. We then have $567^2 + 144^2 = 585^2$ and the equivalent relation $T_{211} + T_{355} = 2T_{292}$. In Table 5.1 we show the solutions of eqn (5.1.10) corresponding to the primitive Pythagorean triples in Table 1.1.

Table 5.1 The triangular form of Pythagoras' theorem.

a	b	c	L	M	$2k$	T_L	T_M	T_{2k}
3	4	5	3	0	2	6	0	3
5	12	13	8	3	6	36	6	21
15	8	17	11	3	8	66	6	36
7	24	25	15	8	12	120	36	78
21	20	29	20	0	14	210	0	105
9	40	41	24	15	20	300	120	210
35	12	37	23	11	18	276	66	171
11	60	61	35	24	30	630	300	465
45	28	53	36	8	26	666	36	351
33	56	65	44	11	32	990	66	528
13	84	85	48	35	42	1176	630	903
63	16	65	39	23	32	780	276	528
55	48	73	51	3	36	1326	6	666
39	80	89	59	20	44	1770	210	990
77	36	85	56	20	42	1596	210	903
65	72	97	68	3	48	2346	6	1176

Exercises 5.1

5.1.1 It is known that, if p is an odd prime of the form $4k + 1$, then there exists an integer x such that $x^2 = -1 \pmod{p}$. What is the corresponding result for triangular numbers?

5.1.2 If the binary operation $*$ is defined by $A * B = 8AB + A + B$ (rather than by eqn (5.1.5)) then prove that $T_l * T_m = T_{2lm+l+m}$. If $x^2 = 8T_l + 1$ and $y^2 = 8T_m + 1$ then find, in terms of T_l and T_m, expressions for x^2y^2 and x^4.

5.1.3 Prove that $(T_1 + T_4) * (T_2 + T_3) = T_{25} + T_{13}$, where the binary operation $*$ is given by eqn (5.1.5). What does this correspond to in terms of squares?

5.1.4 Find T_l and T_m such that $4(T_l + T_m) + 1 = (4k + 3)^2$.

5.1.5 Find a value of l such that $4T_l + 1 = 0 \pmod{53}$.

5.1.6 Find the triangular form of the Pythagorean triple $(15, 112, 113)$.

5.2 More on triangular numbers

In this section we establish a number of theorems concerning the representation of integers as the sum of three or four triangular numbers. We use, without proof, Lagrange's famous theorem that every positive integer may be represented as the sum of no more than four perfect squares, and the theorem that all integers 3 (mod 8) are expressible as the sum of three odd perfect squares. Lagrange's theorem is proved in most books on number theory, such as Niven *et al.* (1991) and Jones and Jones (1998). An account of Gauss' work on integers expressible as the sum of three squares may be found in Rose (1988). An extended account of the work in this section is to be found in Bradley (1988).

Lagrange's four-square theorem asserts (among other things) that the equation

$$x^2 + y^2 + z^2 + w^2 = 2N + 1 \tag{5.2.1}$$

is solvable for integers x, y, z, and w for all positive integers N. The corresponding result for triangular numbers arises from the following theorem.

Theorem 5.2.1 *We have $x^2 + y^2 + z^2 + w^2 = 2N + 1$ if and only if $N = T_m + T_n + T_p + T_q$ for some (possibly negative) integers m, n, p, and q, where one or three of m, n, p, and q are odd.*

Proof It may be verified that the identity stated in the theorem holds when x, y, z, and w and m, n, p, and q are related by the transformation

$$\begin{aligned} m &= \frac{1}{2}(-x + y + z + w - 1)\,, \\ n &= \frac{1}{2}(x - y + z + w - 1)\,, \\ p &= \frac{1}{2}(x + y - z + w - 1)\,, \\ q &= \frac{1}{2}(x + y + z - w - 1)\,. \end{aligned} \tag{5.2.2}$$

Note that, for $x^2 + y^2 + z^2 + w^2$ to be odd, one or three of x, y, z, and w must be odd, and hence, since $m + n + p + q = x + y + z + w - 2$, one or three of m, n, p, and q must also be odd. □

From this theorem it follows that all positive integers are expressible as the sum of four triangular numbers.

The inverse transformation is

$$\begin{aligned} x &= \frac{1}{2}(-m + n + p + q + 1)\,, \\ y &= \frac{1}{2}(m - n + p + q + 1)\,, \\ z &= \frac{1}{2}(m + n - p + q + 1)\,, \\ w &= \frac{1}{2}(m + n + p - q + 1)\,. \end{aligned} \tag{5.2.3}$$

Note that negative values for x, y, z, and w and m, n, p, and q cannot be avoided, but, as $(-x)^2 = x^2$ and $T_{-n-1} = T_n$, the algebra takes care of this without any difficulty.

Note also the dual relationship that allows an inversion of the roles of x, y, z, and w and m, n, p, and q; thus if

$$x^2 + y^2 + z^2 + w^2 = 2(T_m + T_n + T_p + T_q) + 1 \tag{5.2.4}$$

then

$$m^2 + n^2 + p^2 + q^2 = 2(T_{x-1} + T_{y-1} + T_{z-1} + T_{w-1}) + 1\,. \tag{5.2.5}$$

Verification of eqn (5.2.5) is left to the reader.

It is well known that four squares are sufficient to express any positive integer as the sum of perfect squares. They are also necessary since no positive integer of the form $M = 4^L(8N + 7)$, where L and N are integers, can be expressed as the sum of three perfect squares, though all other positive integers can be expressed in this way. It might be thought, in view of this, that some analogous result would hold in expressing a positive integer as the sum of triangular numbers; i.e., that there might be some class of integers for which four triangular numbers would be strictly necessary. However, the surprise is that one can always make do with three. That is, *every positive integer N can be expressed as the sum of at most three triangular numbers*. This result, due to

Gauss, is equivalent to the fact that three perfect squares are sufficient for representing integers of the form $8N + 3$. Thus, the equation $x^2 + y^2 + z^2 = 8N + 3$ possesses nonzero integer solutions x, y, and z for all integers N. Since the right-hand side is 3 (mod 4), it follows that each of x, y, and z is odd, and so there exist integers a, b, and c such that

$$(2a + 1)^2 + (2b + 1)^2 + (2c + 1)^2 = 8N + 3\,, \tag{5.2.6}$$

from which it follows that

$$N = T_a + T_b + T_c\,. \tag{5.2.7}$$

For example, $115 = 9^2 + 5^2 + 3^2$, corresponding to $14 = T_4 + T_2 + T_1$.

This has the following interesting consequence. Given the four-square representation of any odd number:

$$x^2 + y^2 + z^2 + w^2 = 2(T_m + T_n + T_p + T_q) + 1\,, \tag{5.2.8}$$

we now know that three triangular numbers will do, and so one of m, n, p, and q can be chosen to be zero. It follows from the equations of the transformation that in that representation either $y + z + w = x + 1$, or $z + w + x = y + 1$, or $w + x + y = z + 1$, or $x + y + z = w + 1$. Given any one of these possibilities, one may then reverse the sign of one of x, y, z, and w, as appropriate, to obtain a representation in which $x + y + z + w = 1$. We have now proved the following theorem.

Theorem 5.2.2 *Out of all possible solutions in integers of the equation* $x^2 + y^2 + z^2 + w^2 = 2N + 1$, *for any fixed positive integer* N, *it is always possible to find at least one solution satisfying the following additional requirement:* $x + y + z + w = 1$. □

The corresponding theorem on triangular numbers is as follows.

Theorem 5.2.3 *Out of all possible solutions in integers* m, n, p, *and* q *of the equation* $T_m + T_n + T_p + T_q = N$, *for any fixed positive integer* N, *it is always possible to find at least one solution satisfying the following additional requirement:* $m + n + p + q = -1$. □

Referring back to eqns (5.2.2), we see that in these special cases $x = -m$, $y = -n$, $z = -p$, and $w = -q$, and hence we have the surprising dual relationship

$$2N + 1 = m^2 + n^2 + p^2 + q^2 = 2(T_m + T_n + T_p + T_q) + 1\,, \tag{5.2.9}$$

which holds for all positive integers N. For example, with $N = 23$, we have $m = -6$, $n = 3$, $p = 1$, and $q = 1$.

Exercises 5.2

5.2.1 In the transformation of Theorem 5.2.1, find x, y, z, and w when $m = 3$, $n = 2$, $p = 4$, and $q = 6$, and verify the theorem in this case.

5.2.2 Express 87 as the sum of four perfect squares in four different ways, and hence express 43 as the sum of four triangular numbers in three different ways, and as the sum of three triangular numbers in one way.

5.2.3 Find the dual representation of 87.

5.3 Pentagonal and N-gonal numbers

In Sections 5.1 and 5.2 we dealt at some length with triangular numbers. For positive n these numbers can be visualised as consisting of n rows, the top row containing one dot, the second two dots, the third three dots, and so on. For $n = k$, where $k = 2, 3, \ldots$, the array of dots forms a triangle. *Square numbers*, $S_n = n^2$, can be formed by square arrays of dots, and, given the $n \times n$ array, a border of $2n + 1$ dots around two sides converts it into the $(n+1) \times (n+1)$ square array. This is, of course, because $n^2 + (2n + 1) = (n + 1)^2$. S_5 is illustrated in Fig. 5.2. It also shows that a square number is the sum of two consecutive triangular numbers.

The pentagonal numbers

Pentagonal numbers are defined by the equation

$$P_n = \frac{1}{2}n(3n - 1)\,, \tag{5.3.1}$$

from which $P_1 = 1$, $P_2 = 5$, $P_3 = 12$, $P_4 = 22$, $P_5 = 35$, etc. P_2 can be visualised by five dots in the shape of a pentagon. To visualise P_3, take the pentagon from the P_2 diagram and extend two of its sides so that they have three dots, rather than two. Then complete the new pentagon so that it has three dots on each side. The resulting figure has an outer pentagon with ten dots, but it has two dots from the original pentagon internal to it. This building process is then continued to form an outer pentagon that has

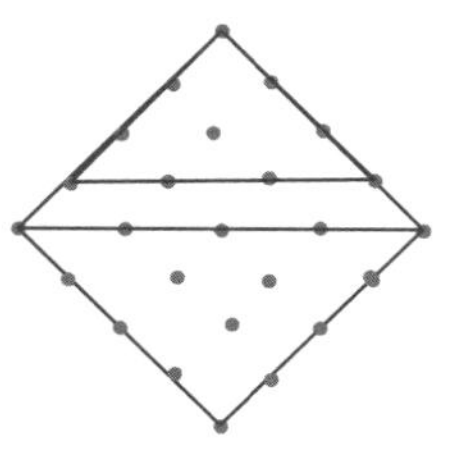

Fig. 5.2 The square number $S_5 = T_4 + T_5 = 10 + 15 = 25$.

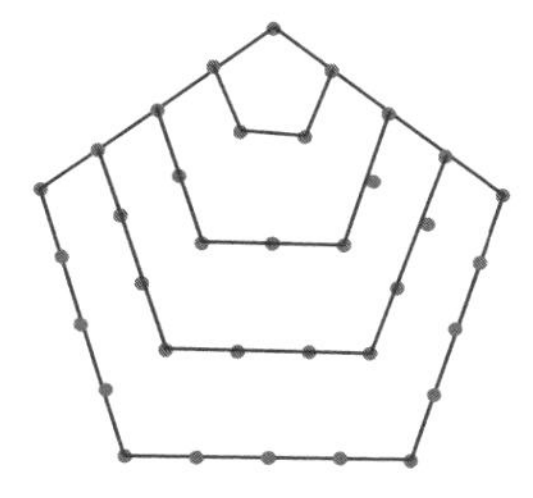

Fig. 5.3 The pentagonal number $P_5 = 35$.

four dots on each side, and so on, see Fig. 5.3. To go from P_n to P_{n+1} we have to add a border of $3n + 1$ dots, so the formula for P_n, given by eqn (5.3.1), may be established by induction, since $\frac{1}{2}n(3n - 1) + 3n + 1 = \frac{1}{2}(n + 1)(3n + 2)$. Every pentagonal number is one-third of a triangular number. To be precise, $P_n = (1/3)T_{3n-1}$.

N-gonal numbers for $N > 5$

It may be observed that one moves from T_n to T_{n+1} by adding a border of $n + 1$ dots, that one moves from S_n to S_{n+1} by adding a border of $2n+1$ dots, and that one moves from P_n to P_{n+1} by adding a border of $3n + 1$ dots.

Generalising, we may define N-gonal numbers by the recurrence relation $N_1 = 1$ and

$$N_{n+1} = N_n + (N - 2)n + 1\,. \tag{5.3.2}$$

It may now be shown, by induction, that

$$N_n = \frac{1}{2}n\left[(N - 2)n + (4 - N)\right]. \tag{5.3.3}$$

Note, as a check, that this agrees with T_n, S_n, and P_n for $N = 3, 4$, and 5, respectively. The case $N = 6$, $n = 5$ is shown in Fig. 5.4.

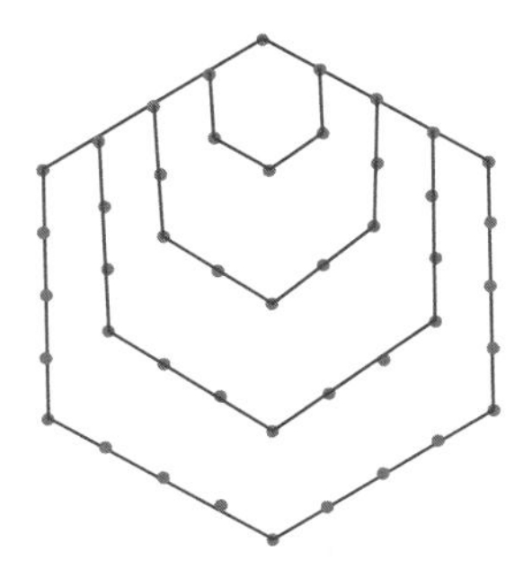

Fig. 5.4 The hexagonal number $H_5 = 45$.

N-gonal numbers that are square for $N > 2$

For an N-gonal number to be a square number there must be a positive integer x such that

$$(N-2)n^2 - (N-4)n = 2x^2\,, \tag{5.3.4}$$

which, on multiplying by $4(N-2)$ and completing the square, reduces to

$$[2(N-2)n - (N-4)]^2 - 8(N-2)x^2 = (N-4)^2\,. \tag{5.3.5}$$

(For $N = 4$, this reduces, as it must, to $x = n$, since all 4-gonal numbers are square numbers, by definition.) We are assured of one solution of eqn (5.3.5) for each N. This is because $x = 1$ when $n = 1$, whatever the value of N.

We consider first the case when $8(N-2)$ is a perfect square, that is, when $N = 2 + 2k^2$ for some integer k. Equation (5.3.5) then becomes

$$[2k^2n - (k^2-1)]^2 - 4k^2x^2 = (k^2-1)^2\,. \tag{5.3.6}$$

For example, when $N = 20$, we have $k = 3$ and eqn (5.3.6) becomes $(9n + 3x - 4)(9n - 3x - 4) = 16$. This equation has only one positive integral solution $n = 1$ and $x = 1$. So, when $N = 20$, only N_1 is square. When $N = 34$, we have $k = 4$ and eqn (5.3.6) becomes $(32n - 15 + 8x)(32n - 15 - 8x) = 225$, and this equation has the two positive integral solutions $n = 1$ and $x = 1$, and $n = 4$ and $x = 14$. In general, when $N = 2k^2 + 2$ and $k > 1$, then, since $(k^2-1)^2$ has only a finite number of pairs of factors in positive integers, only a finite number of such numbers can be square numbers. We have proved the following theorem.

Theorem 5.3.1 *When $N = 2k^2 + 2$ ($k > 1$), where k is an integer, only a finite number of N-gonal numbers are perfect squares.* □

The case $N = 7$

We consider whether there are any heptagonal numbers, other than 1, that are perfect squares. If there are, then, by putting $N = 7$ in eqn (5.3.5), there must be solutions other than $y = 7$, $n = 1$, and $x = 1$ of the Pell-like equation

$$y^2 - 40x^2 = 9\,, \tag{5.3.7}$$

where $y = 10n - 3$.

The basic solutions (those that cannot be derived from one another by iteration) are $y = 7$ and $x = 1$ corresponding to $n = 1$, $y = 57$ and $x = 9$ corresponding to $n = 6$, and $y = 487$ and $x = 77$ corresponding to $n = 49$. In order to generate other solutions we require the basic solution of the auxiliary equation $y^2 - 40x^2 = 1$, in which $y = 1 \pmod{10}$. This is $y = 721$ and $x = 114$. (The smaller solution with $y = 19$ and $x = 3$ will not do, as it has $y = -1 \pmod{10}$ and it would not then produce solutions of the original equation that are equal to 7 (mod 10).) The equations for generating solutions are given by

$$y_{n+1} = 721y_n + 4560x_n\,, \quad x_{n+1} = 114y_n + 721x_n\,, \tag{5.3.8}$$

with the first sequence starting with $y_1 = 7$ and $x_1 = 1$. This gives $y_2 = 9607$ and $x_2 = 1519$. There are an infinite number of heptagonal numbers that are squares, generated by the above recurrence relations. The first four are $n = 1$ and $N_1 = 1$, $n = 6$ and $N_6 = 81 = 9^2$, $n = 49$ and $N_{49} = 5929 = 77^2$, and $n = 961$ and $N_{961} = 2\,307\,361 = 1519^2$.

The case $N = 8$

We consider whether there are any octagonal numbers, other than 1, that are perfect squares. If there are, then, by putting $N = 8$ in eqn (5.3.5), there must be solutions other than $y = 4$, $n = 1$, and $x = 1$ of the Pell-like equation

$$y^2 - 12x^2 = 4\,, \tag{5.3.9}$$

where $y = 6n - 2$.

The basic solution is $y = 4$ and $x = 1$ corresponding to $n = 1$. In order to generate other solutions we determine the smallest solution of the auxiliary equation $y^2 - 12x^2 = 1$ in which $y = 1$ (mod 6). This is $y = 7$ and $x = 2$, and so the recurrence relations for providing solutions are

$$y_{n+1} = 7y_n + 24x_n\,, \quad x_{n+1} = 2y_n + 7x_n\,. \tag{5.3.10}$$

With $y_1 = 4$ and $x_1 = 1$ we find that $y_2 = 52$ and $x_2 = 15$, corresponding to $n = 9$. Thus the ninth octagonal number is the perfect square $225 = 15^2$. The next octagonal number to be a perfect square has $n = 121$ and has the value $43\,681 = 209^2$.

N-gonal numbers that are also triangular numbers

The equation $N_n = T_m$ becomes, after completing the square,

$$y^2 - (N-2)x^2 = (N-6)(N-3)\,, \tag{5.3.11}$$

where $y = 2(N-2)n - (N-4)$ and $x = 2m + 1$. This verifies that $N = 3$ and $N = 6$ are special cases in which all the N-gonal numbers are triangular, see Exercise 5.3.2. Only a finite number of N-gonal numbers are triangular when $N = k^2 + 2$, $k > 2$, and eqn (5.3.11) becomes

$$y^2 - k^2x^2 = (k^2-4)(k^2-1)\,, \tag{5.3.12}$$

where $y = 2k^2n - (k^2 - 2)$. This is because the term on the right-hand side of eqn (5.3.12) has only a finite number of factors.

For example, when $N = 18$, we have $k = 4$ and eqn (5.3.12) becomes

$$(y - 4x)(y + 4x) = 180\,, \tag{5.3.13}$$

with $y = 32n - 14$ and $x = 2m + 1$; this is satisfied for positive m and n by $y = 18$ and $x = 3$ only, showing that only the trivial solution with $n = 1$ exists.

For $N = 5$ eqn (5.3.11) becomes

$$y^2 - 3x^2 = -2\,, \tag{5.3.14}$$

where $y = 6n - 1$ and $x = 2m + 1$.

The basic solution is $y = 5$, $x = 3$, $n = 1$, and $m = 1$, and the next smallest solution is $y = 71$, $x = 41$, $n = 12$, and $m = 20$, the common value being 210. Successive solutions are given by $y' = 7y + 12x$ and $x' = 4y + 7x$, where we note, with satisfaction, that if x is odd then so is x', and if $y = -1 \pmod 6$ then so is y'.

Hexagonal close-packed numbers

These are the numbers Hex_n represented by circles in close-packed hexagon form, which therefore satisfy the relations $\mathrm{Hex}_1 = 1$ and $\mathrm{Hex}_{n+1} = \mathrm{Hex}_n + 6n$. The formula for them in terms of n is

$$\mathrm{Hex}_n = 3n^2 - 3n + 1\,. \tag{5.3.15}$$

The first four are 1, 7, 19, and 37. The first, eighth, and 105th Hex numbers are the perfect squares 1, 169, and 32 761, respectively. Note that $n^3 - (n-1)^3 = 3n^2 - 3n + 1 = \mathrm{Hex}_n$, so that

$$\mathrm{Hex}_1 + \mathrm{Hex}_2 + \mathrm{Hex}_3 + \ldots + \mathrm{Hex}_n = n^3\,. \tag{5.3.16}$$

Hex_3 is shown in Fig. 5.5.

Centred square numbers

These are the numbers C_n represented by circles in close-packed square form, which therefore satisfy the relations $C_1 = 1$ and $C_{n+1} = C_n + 4n$. The formula for them in terms of n is

$$C_n = 2n^2 - 2n + 1\,. \tag{5.3.17}$$

The first four are 1, 5, 13, and 25. The first, fourth, and twenty-first centred square numbers are the perfect squares 1, 25, and 841, respectively.

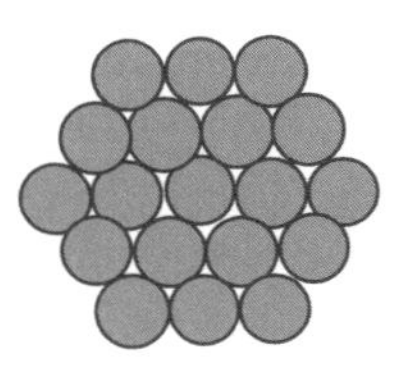

Fig. 5.5 The hexagonal close-packed number $\mathrm{Hex}_3 = 19$.

Exercises 5.3

5.3.1 Find the smallest n, $n > 1$, for which the pentagonal number P_n is a square number.

5.3.2 Show that all hexagonal numbers ($N = 6$) are triangular numbers.

5.3.3 Find all 74-gonal numbers that are perfect squares.

5.3.4 Prove that $(N+1)_n - N_n = T_{n-1}$.

5.3.5 Find the smallest non-trivial nonagonal number ($N = 9$) that is a perfect square.

5.3.6 Prove that, if N_n is an octagonal number that is also a perfect square, then n is a perfect square.

5.3.7 Find the smallest hexagonal close-packed number after 32 761 that is a perfect square.

5.3.8 Find the smallest centred square number after 841 that is a perfect square.

5.4 Polyhedral numbers

Cubes

In Section 5.3 we showed that the sum of the first n Hex numbers is the cube n^3. Another remarkable and well-known property of the cubes is that

$$1^3 + 2^3 + 3^3 + \ldots + n^3 = (1 + 2 + 3 + \ldots + n)^2 = T_n^2 \,. \tag{5.4.1}$$

We write $\mathrm{Cub}_n = n^3$, and such numbers are illustrated in Fig. 5.6.

Tetrahedral numbers

Tetrahedral numbers arise when the triangular numbers are thought of as spheres in triangular arrays that are placed in layers on top of one another in the shape of a pyramid. Thus

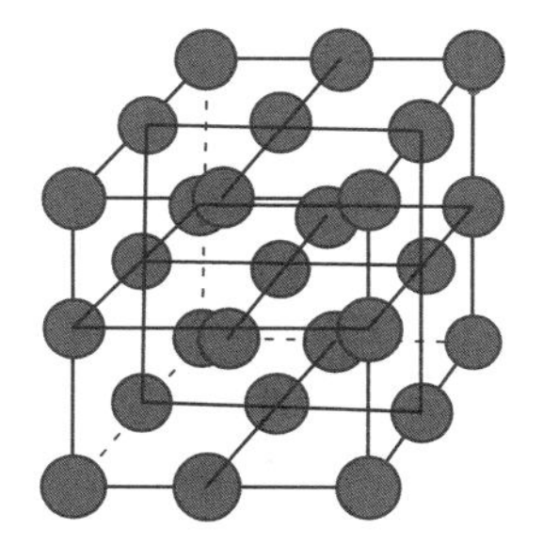

Fig. 5.6 The cubes $\mathrm{Cub}_3 = 27$.

$$\mathrm{Tet}_n = T_1 + T_2 + T_3 + \ldots + T_n = \frac{1}{6}n(n+1)(n+2)\,. \tag{5.4.2}$$

The first four are 1, 4, 10, and 20. The elliptic curve with equation

$$6y^2 = x(x+1)(x+2) \tag{5.4.3}$$

has integer points $(-2,0)$, $(-1,0)$, $(0,0)$, $(1,1)$, $(1,-1)$, $(2,2)$, $(2,-2)$, $(48,140)$, and $(48,-140)$, and, as confirmed by Kevin Buzzard (private communication), these are the only integer points. Hence only the first, second, and forty-eighth tetrahedral numbers are perfect squares. Tet_4 is shown in Fig. 5.7.

Square pyramid numbers

Square pyramid numbers arise when the square numbers are thought of as spheres in square arrays that are placed in layers on top of one another in the shape of a pyramid. Thus

$$\mathrm{Pyr}_n = 1^2 + 2^2 + 3^2 + \ldots + n^2 = \frac{1}{6}n(n+1)(2n+1)\,. \tag{5.4.4}$$

The first four are 1, 5, 14, and 30. Only the first and twenty-fourth square pyramid numbers are perfect squares. Pyr_3 is shown in Fig. 5.8.

Octahedral numbers

Octahedral numbers arise when a square pyramid of n layers is stuck to an inverted square pyramid of $n-1$ layers. Thus

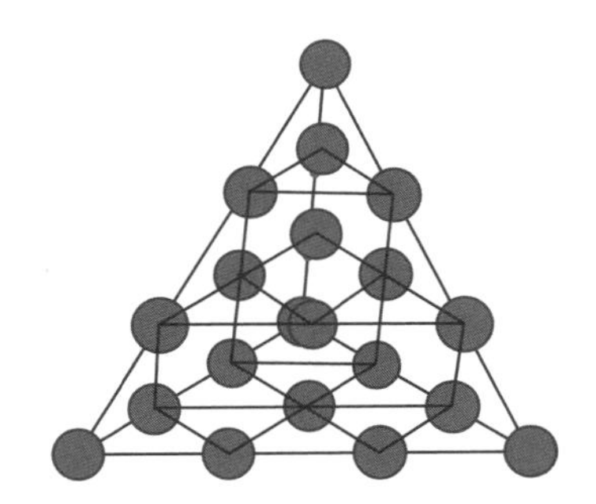

Fig. 5.7 The tetrahedral number $\mathrm{Tet}_4 = 20$.

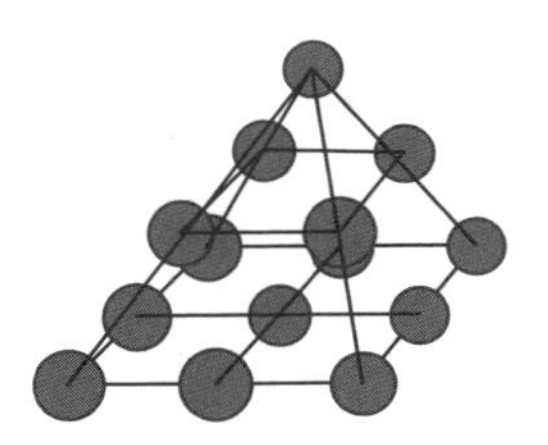

Fig. 5.8 The square pyramid number $\mathrm{Pyr}_3 = 14$.

$$\begin{aligned}\mathrm{Oct}_n &= \mathrm{Pyr}_n + \mathrm{Pyr}_{n-1} \\ &= \frac{1}{6}n(n+1)(2n+1) + \frac{1}{6}n(n-1)(2n-1) \\ &= \frac{1}{3}n(2n^2+1)\,.\end{aligned} \tag{5.4.5}$$

The first four are 1, 6, 19, and 44, see Fig. 5.9.

Body-centred cubic numbers

Body-centred cubic numbers arise when two cubes occupy alternate layers with their vertices at the centres of the alternate cube. Thus

$$\mathrm{Bcc}_n = (n-1)^3 + n^3 = (2n-1)(n^2-n+1)\,. \tag{5.4.6}$$

The first four are 1, 9, 35, and 91. The only positive integers for which Bcc_n is a perfect square are $n = 1$ and 2. There are only trivial solutions ($n = 1$ and $n = 2$) to the problem of whether the sum of two consecutive cubes is a perfect square, see Fig. 5.10.

Rhombic dodecahedral numbers

Rhombic dodecahedral numbers are the analogue in three dimensions of the hexagonal close-packed numbers. These were given by $\mathrm{Hex}_n = (n+1)^3 - n^3$, and so

$$\mathrm{Rho}_n = (n+1)^4 - n^4\,. \tag{5.4.7}$$

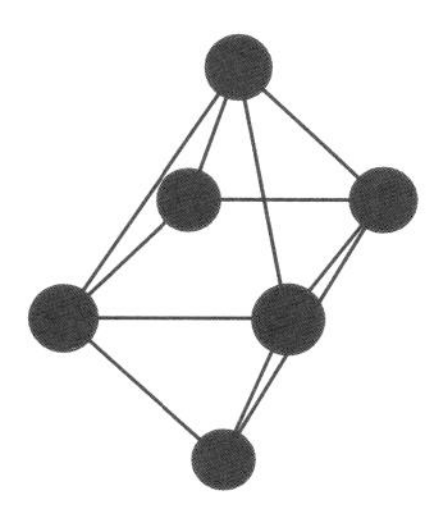

Fig. 5.9 The octahedral number $\mathrm{Oct}_2 = 6$.

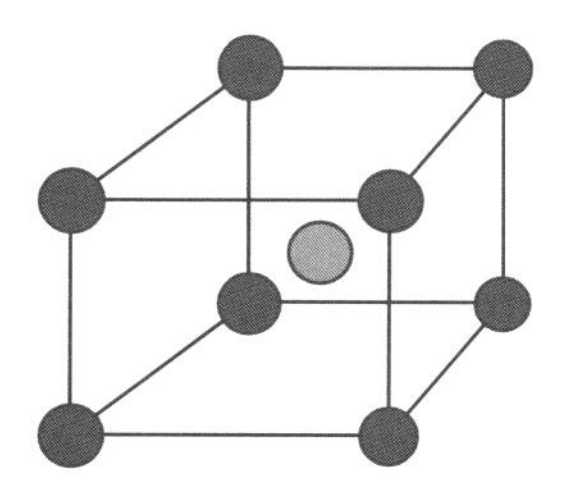

Fig. 5.10 The body-centred cubic number $\mathrm{Bcc}_2 = 9$.

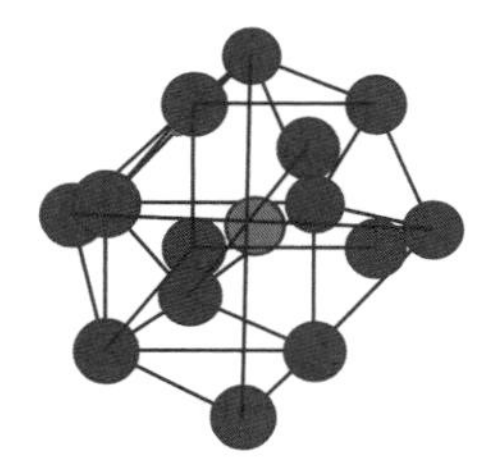

Fig. 5.11 The rhombic dodecahedral number $\mathrm{Rho}_2 = 15$.

The first four are 1, 15, 65, and 175, see Fig. 5.11.

Exercise 5.4

5.4.1 Is the difference between two consecutive cubes ever a perfect square?

5.5 Catalan numbers

The nth Catalan number is

$$\mathrm{Cat}_n = \left(\frac{1}{n+1}\right) {}^{2n}C_n = \frac{(2n)!}{(n+1)!n!}. \tag{5.5.1}$$

The first few are $\mathrm{Cat}_0 = 1$, $\mathrm{Cat}_1 = 1$, $\mathrm{Cat}_2 = 2$, $\mathrm{Cat}_3 = 5$, $\mathrm{Cat}_4 = 14$, and $\mathrm{Cat}_5 = 42$. From the formula for Cat_n it is not even clear that it is an integer for all non-negative n. This will become clear at the end of this section. It turns out that the Catalan number is the solution to several counting problems of a geometrical nature.

Non-crossing segments joining pairs of points of a regular $2n$-gon

The problem chosen to illustrate the Catalan numbers is concerned with line segments in a regular $2n$-gon. The vertices are labelled consecutively in a clockwise (or anti-clockwise) order around the polygon. The counting rules are (i) that every vertex must be joined by a line segment to another vertex, and (ii) that no two line segments should cross one another. The problem is to count the number V_n of such configurations. To be clear, we need to stipulate when two configurations are to be classified as distinct, see Fig. 5.12. You will observe from the figure that we count two configurations as being distinct when one may be rotated into another. So, in the case $n = 3$, the configuration 1, 2; 3, 4; 5, 6 is distinct from the configuration 2, 3; 4, 5; 6, 1. In other words, the labelling matters.

Suppose that we have such a $2n$-gon. It is clear that vertex 1 must be joined to a vertex with an even label. We count the number of figures in which vertex 1 is joined

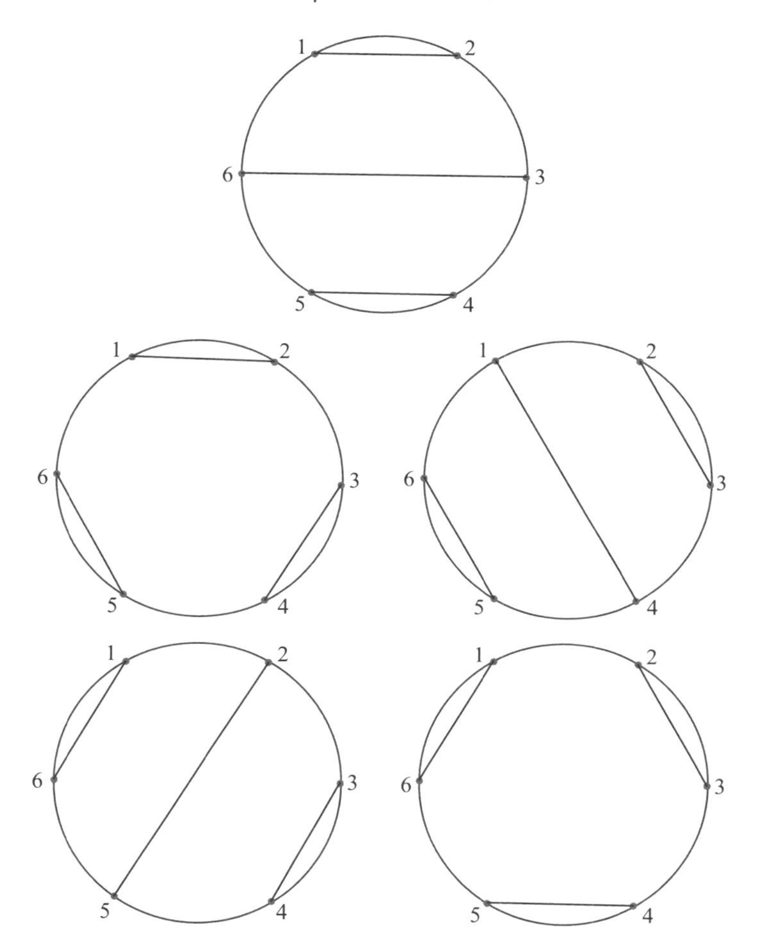

Fig. 5.12 The Catalan number $\mathrm{Cat}_3 = 5$.

to vertex $2m$ by a line segment ($1 \leqslant m \leqslant n$). This line divides the figure up into two parts. On one side there are the vertices $2, 3, \ldots, 2m-1$ and on the other side there are the vertices $2m+1, 2m+2, \ldots, 2n$. The number of vertices in the first part is $2m-2$, and the number of vertices in the second part is $2n-2m$. So the number of ways that these figures can be filled up with line segments is V_{m-1} for the first part and V_{n-m} for the second part. The number of configurations with vertex 1 joined to vertex $2m$ is therefore $V_{m-1}V_{n-m}$. It follows, by varying the vertex $2m$, that

$$V_n = V_0 V_{n-1} + V_1 V_{n-2} + \ldots + V_{m-1} V_{n-m} + \ldots + V_{n-1} V_0 \,, \tag{5.5.2}$$

where, for consistency, we define $V_0 = 1$.

In order to solve the recurrence relation (5.5.2), we define the generating function

$$F(x) = V_0 x + V_1 x^2 + V_2 x^3 + \ldots + V_n x^{n+1} + \ldots \,. \tag{5.5.3}$$

Then

$$[F(x)]^2 = V_0V_0x^2 + (V_0V_1 + V_1V_0)x^3 + (V_0V_2 + V_1V_1 + V_2V_0)x^4 + \dots , \quad (5.5.4)$$

and by eqn (5.5.2) this is equal to $F(x) - x$. Solving for $F(x)$, we obtain

$$F(x) = \frac{1}{2}\left(1 - \sqrt{1-4x}\right), \quad (5.5.5)$$

where the minus sign has been chosen since $F(0) = 0$. The coefficient of x^{n+1} in the binomial expansion of this is $\frac{1}{2}(1/2)(1/2)(3/2)\cdots((2n-1)/2)(1/(n+1)!)(4^{n+1})$, since all of the minus signs cancel. This expression is equal to $\frac{1}{2}[1\times 3\times 5\times\cdots\times(2n-1)\times 2^{n+1}]/(n+1)! = 2n!/(n+1)!n! = \mathrm{Cat}_n$. Since V_n is an integer, it follows that Cat_n is an integer. Cat_n is also the answer to other geometrical counting problems, see the exercise below and the entertaining book by Conway and Guy (1996).

Exercise 5.5

5.5.1 Prove that Cat_n is also the number of distinct ways that an $(n+2)$-gon can be split up by diagonals into n non-overlapping triangles, with the rule that the vertices are labelled, so that figures that are obtained by rotation from one another are counted as distinct.

6 Quadrilaterals and triangles

The first problem that we consider in this chapter is the analogue for parallelograms of the Heron triangle. That is, we consider when a parallelogram has integer sides, integer diagonals, and integer area. Since such parallelograms are the union of two triangles with integer sides and integer or half odd integer area, the difficulty lies in making the second diagonal an integer. Solutions in particular cases are given below, but a general solution seems unlikely. Since the problem is trivial when the parallelogram is a rectangle, we exclude this case from consideration.

We then look, once again, at the cyclic quadrilateral and consider the problem of how to choose the sides of a cyclic quadrilateral so that it has both integer sides and integer area. There is the more general problem of finding those cyclic quadrilaterals that have integer diagonals as well, but the solution to that problem has recently been given by Sastry (2003) and is not repeated here.

The next problem is concerned with a triangle ABC having integer sides, together with a point P in the plane of the triangle, not on the sides. From P, perpendiculars to the sides are drawn, meeting BC, CA, and AB at L, M, and N, respectively. It is shown how to choose P and the side lengths of ABC so that the six line segments BL, LC, CM, MA, AN, and NB are all of integer length. It is also shown how to locate P when the six line segments have given integer lengths.

The final arithmetical problem in this chapter turns out to have two equivalent geometrical presentations. The first is the problem of whether, given an integer-sided equilateral triangle, there are points in the plane of the triangle that are at a rational distance from each vertex. This problem is solved in that we establish that there an infinite number of points having this property which are (i) on the sides of the triangle, (ii) on the circumcircle of the triangle, and (iii) internal to the triangle. Moreover, it is evident that all of the points lie on certain lines through the vertices of the triangle. The second problem is concerned with an integer-sided triangle ABC, all of whose angles are less than 120°, and its Fermat point F. The problem is to find those integer-sided triangles for which all three distances AF, BF, and CF are of integer length. These two problems turn out to be the same in the sense that the two solutions both involve the same three quadratic forms, all of which have to be made perfect squares.

6.1 Integer parallelograms

We define an *integer parallelogram* $ABCD$ to be one with integer sides, integer diagonals, and integer area. We exclude rectangles and therefore insist that the parallelogram possesses an acute angle. We first provide an account of the problem from an algebraic point of view. Then we consider two ways that are sufficient to ensure that a parallelogram is an integer parallelogram, and list a number of special cases. However, as the general solution depends on making a polynomial of the eighth degree in four variables a perfect square, it is highly unlikely that a general solution can be written down.

If the sides are of length a and b, with $a \geqslant b$, the acute angle between them is θ, and the diagonals are of length c and d, with $d > c$, then by the cosine rule $a^2 + b^2 - 2ab\cos\theta = c^2$ and $a^2 + b^2 + 2ab\cos\theta = d^2$, so that

$$2(a^2 + b^2) = c^2 + d^2 \tag{6.1.1}$$

and

$$\cos\theta = \frac{d^2 - c^2}{4ab}. \tag{6.1.2}$$

The problem of finding a general solution in integers of eqn (6.1.1) is easy enough, but for the area $[ABCD]$ to be integral we have to ensure that $[ABCD] = ab\sin\theta$ is an integer. This means that the parameters p, q, r, and s, defined below, from which a, b, c, and d are obtained, have to be carefully chosen so that when $\cos\theta$ is found from eqn (6.1.2) then the corresponding value of $\sin\theta = \sqrt{1 - \cos^2\theta}$ is rational. A general method for achieving this is unlikely; but it is possible, in certain cases, to make $\sin\theta$ rational, and we show, by example, two ways of proceeding.

From eqn (6.1.1) we have $(a+b)^2 + (a-b)^2 = c^2 + d^2$, so there exist integers p, q, r, and s such that $a + b = pq + rs$, $a - b = pr - qs$, $c = pq - rs$, and $d = pr + qs$. Note that if a and b are both even then $c^2 + d^2 = 0 \pmod 4$, and hence c and d are both even. It is then possible to divide all of them by 2. If we insist that a, b, c, and d have no common factor 2, then **either** we must choose a and b to be of opposite parity, in which case c and d must both be odd, **or** we must choose a and b to both be odd, in which case c and d must both be even. These parity considerations limit the choices that may be made for the parameters p, q, r, and s. Note also that for $a \geqslant b$ and $d > c > 0$ we need to have $pr \geqslant qs$, $pq > rs$, and $pr + qs + rs > pq$. Substituting the values for a, b, c, and d in terms of the parameters p, q, r, and s into eqn (6.1.2), we find that

$$\cos\theta = \frac{A - B}{A + B}, \tag{6.1.3}$$

where $A = 4pqrs$ and $B = (p^2 - s^2)(q^2 - r^2)$. In order for $\sin\theta$ to be rational it is necessary, since

$$\sin^2\theta = \frac{4AB}{(A+B)^2}, \tag{6.1.4}$$

for

$$4AB = 16pqrs(p^2 - s^2)(q^2 - r^2) \tag{6.1.5}$$

to be a perfect square.

Since $0 \leqslant \cos\theta < 1$, we must choose $p > s$ and $q > r$, as well as having $pr \geqslant qs$ and $pr + qs + rs > pq$.

Example 6.1.1 The first and most obvious method for ensuring that $pqrs(p^2 - s^2)(q^2 - r^2)$ is a perfect square is to find values of the parameters so that $ps(p^2 - s^2) = qr(q^2 - r^2)$. For example, $p = 7$, $s = 3$, $q = 7$, and $r = 5$; this gives $a = 39$, $b = 25$, $c = 34$, and $d = 56$, and $\cos\theta = 33/65$ and $\sin\theta = 56/65$, see Fig. 6.1. We leave it as an exercise for the reader to show that there are an infinite number of possibilities of this type, see Section 1.2.

There are also an infinite number of possibilities to be found by putting $p = q$ and then either $r = s$ or $r^2 + rs + s^2 = p^2$.

When $pr = qs$ the parallelogram is a rhombus. The expression $pqrs(p^2 - s^2)(q^2 - r^2)$ is then automatically a perfect square and a corresponding solution always exists. Integer solutions of the equation $r^2 + rs + s^2 = p^2$ are those studied in Section 1.2 in connection with integer-sided triangles having an angle of 120°. From Table 1.2 the four lowest possible values of p are 7, 13, 19, and 21.

Example 6.1.2 We now give another method for ensuring that $pqrs(p^2 - s^2)(q^2 - r^2)$ is a perfect square. This can be achieved by setting $p = u^2 + v^2$, $s = 2uv$, $q = h^2 + k^2$, and $r = 2hk$, where u and v, and h and k are coprime and of opposite parity. Then $p^2 - s^2 = (u^2 - v^2)^2$ and $q^2 - r^2 = (h^2 - k^2)^2$, and it remains to make $uvhk(u^2 + v^2)(h^2 + k^2)$ a perfect square. This may be arranged by choosing u and v, and h and k to be the first two members of a pair of Pythagorean triples in which $uv = hk$. Such pairs of triples certainly exist. For example, $u = 35$, $v = 12$, $h = 21$, and $k = 20$ has $uv = hk = 420$, $u^2 + v^2 = 37^2$, and $h^2 + k^2 = 29^2$. To satisfy parity conditions we actually have to double p and s, giving $p = 2738$, $q = 841$, $s = 1680$, and $r = 840$. This gives an integer parallelogram with $a = 2\,300\,449$, $b = 1\,413\,409$, $c = 891\,458$, $d = 3\,712\,800$, $\cos\theta = 3\,247\,546\,618\,559/3\,251\,475\,320\,641$, and $\sin\theta = 159\,789\,614\,880/3\,251\,475\,320\,641$. Finally, $[ABCD] = 159\,789\,614\,880$. Solutions having such large values for a, b, c, and d are less satisfactory, but the method shows that solutions exist other than the special cases and those similar to the one described in Example 6.1.1.

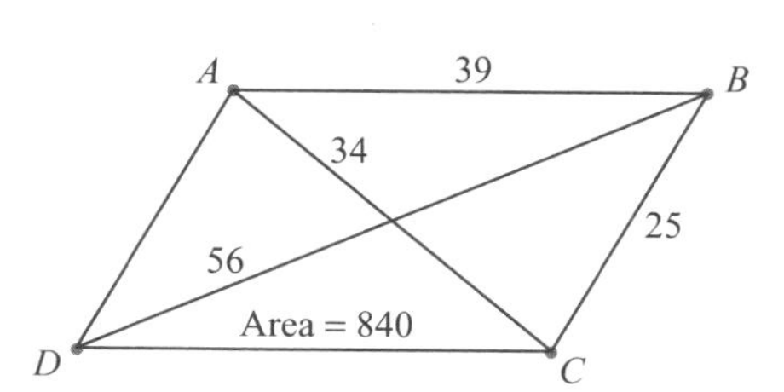

Fig. 6.1 An integer parallelogram.

Exercises 6.1

6.1.1 Repeat Example 6.1.1 with $p = q = 37$ and $r^2 + rs + s^2 = 1369$.

6.1.2 Repeat Example 6.1.2 with $u = 140$, $v = 1221$, $h = 660$, and $k = 259$.

6.2 Area of a cyclic quadrilateral

The aim of this section is to show how to choose a cyclic quadrilateral so that it has integer sides and integer area. Since the problem is trivial for rectangles, we do not consider them further. Nor do we consider cases, except for the isosceles trapezium, in which the diagonals are of integer length and the circumscribing circle has integer radius.

A cyclic quadrilateral with integer sides, integer diagonals, and integer area was called by Sastry (2003) a *Brahmagupta quadrilateral*, and these are characterised in Sastry (2003), where reference is made to a previous, more complex characterisation by Dickson (1971). I had intended to address the more complicated problem, but the article by Sastry (2003) leaves no room for improvement.

The reason Sastry chose the name Brahmagupta quadrilateral was because of the famous formula Brahmagupta established for the area of a cyclic quadrilateral $ABCD$ in terms of its side lengths, a formula similar to Heron's formula for the area of a triangle. This formula is

$$[ABCD] = \frac{1}{4}\sqrt{(b+c+d-a)(c+d+a-b)(d+a+b-c)(a+b+c-d)}\,, \tag{6.2.1}$$

where $AB = a$, $BC = b$, $CD = c$, and $DA = d$. We now give a derivation of this formula.

If $a = c$ and $b = d$, then $ABCD$, being cyclic, must be a rectangle, and Brahmagupta's formula gives the correct value ab for $[ABCD]$. Provided that $ABCD$ is not a rectangle, we may assume without loss of generality, since the formula to be obtained is symmetric in a, b, c, and d, that BA meets CD at P and that P lies on the extension of BA beyond A and on the extension of CD beyond D.

Let $PB = x$ and $PC = y$, so that $PA = x - a$ and $PD = y - c$. Now, since $ABCD$ is cyclic, $\angle PAD = \angle PCB$. It follows that the triangles PAD and PCB are similar. By Heron's formula for the triangle PCB, we have

$$[PCB] = \frac{1}{4}\sqrt{(x+y+b)(y+b-x)(b+x-y)(x+y-b)} \tag{6.2.2}$$

and, because of their similarity, $[PAD] = (d^2/b^2)[PCB]$. Now

$$[ABCD] = [PCB] - [PAD] = \frac{b^2 - d^2}{b^2}[PCB]\,. \tag{6.2.3}$$

Since PAD and PCB are similar, we have $(y-c)/d = x/b$ and $(x-a)/d = y/b$. By addition and subtraction, we obtain $x+y = b(a+c)/(b-d)$ and $x-y = b(a-c)/(b+d)$. It follows that

$$\begin{aligned} x+y+b &= \frac{b(a+b+c-d)}{b-d}\,, \\ x+y-b &= \frac{b(c+d+a-b)}{b-d}\,, \\ b+x-y &= \frac{b(d+a+b-c)}{b+d}\,, \\ y+b-x &= \frac{b(b+c+d-a)}{b+d}\,. \end{aligned} \tag{6.2.4}$$

Substituting these expressions back into Heron's formula for $[PCB]$, and using the relation (6.2.3) between $[ABCD]$ and $[PCB]$, we obtain Brahmagupta's formula.

In order to find a quadrilateral with integer sides and integer area we choose fourteen integer parameters and set

$$\begin{aligned} b+c+d-a &= x^2m^2n^2p^2qrs = f\,, \\ c+d+a-b &= y^2l^2n^2p^2qtu \ \ = g\,, \\ d+a+b-c &= z^2l^2m^2p^2rtv \ \ = h\,, \\ a+b+c-d &= w^2l^2m^2n^2suv = k\,. \end{aligned} \tag{6.2.5}$$

Then

$$\begin{aligned} a &= \frac{1}{4}(g+h+k-f)\,, \\ b &= \frac{1}{4}(h+k+f-g)\,, \\ c &= \frac{1}{4}(k+f+g-h)\,, \\ d &= \frac{1}{4}(f+g+h-k)\,, \end{aligned} \tag{6.2.6}$$

and

$$[ABCD] = \frac{1}{4}l^3m^3n^3p^3xyzwqrstuv\,. \tag{6.2.7}$$

The sides may need to be multiplied by 2 or 4 to ensure that they are integers.

As a first example, with $x = y = z = l = p = r = v = s = u = 1$, $w = m = n = t = 2$, and $q = 3$, we obtain $f = 48$, $g = 24$, $h = 8$, $k = 64$, $a = 12$, $b = 24$, $c = 32$, $d = 4$, and $[ABCD] = 192$. Reducing the sides by a factor of 4 gives $a = 3$, $b = 6$, $c = 8$, $d = 1$, and $[ABC] = 12$, see Fig. 6.2.

As a second example, with $w = l = p = q = r = s = m = n = t = 1$, $x = 6$, $y = 3$, $z = 2$, $u = 2$, and $v = 13$, we obtain $f = 36$, $g = 18$, $h = 52$, $k = 26$, $a = 15$, $b = 24$, $c = 7$, $d = 20$, and $[ABCD] = 234$.

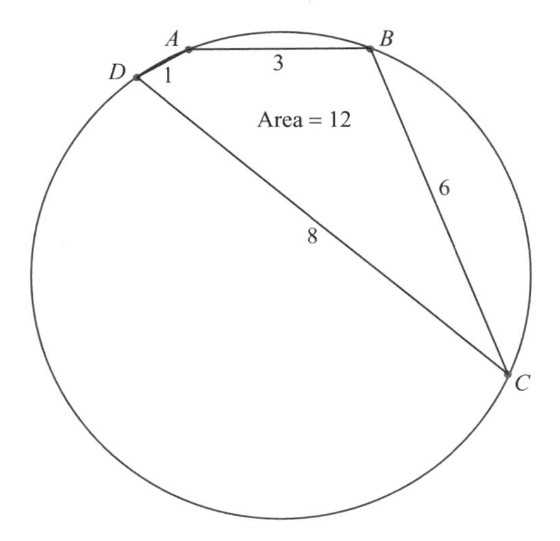

Fig. 6.2 A cyclic quadrilateral with integer sides and integer area.

It must be appreciated that Brahmagupta's formula for the area in terms of the sides is valid only for a cyclic quadrilateral, and not for a general convex quadrilateral. In general, if four loosely-jointed rods of lengths a, b, c, and d satisfying $b+c+d>a$, $c+d+a>b$, $d+a+b>c$, and $a+b+c>d$ are given, then there is only one configuration in which they will settle into that of a cyclic quadrilateral; see Exercise 2.3.2, where you are required to show that the diagonals e and f of a cyclic quadrilateral are uniquely determined by its side lengths.

Isosceles trapezium

An isosceles trapezium is a trapezium in which there is a line of symmetry perpendicular to the two parallel lines. Such a figure is always cyclic. We denote the length of the smaller of the two parallel lines by a, the length of the longer one by c, and the length of the equal legs by b. Then the formula for the area is

$$[ABCD] = \frac{1}{2}(a+c)\sqrt{b^2 - \frac{1}{4}(c-a)^2}\,. \tag{6.2.8}$$

The term under the square root sign is the square of the distance between the parallel lines. It is an easy problem to choose integer side lengths so that the area is also an integer. Every Pythagorean triple provides a solution. The example that follows comes from the triple $(8, 15, 17)$ in an obvious way. Thus, if $a=5$, $b=17$, and $c=35$ then we have $[ABCD] = 20 \times 8 = 160$, see Fig. 6.3.

Isosceles trapezium with rational circumradius

We use the notation that the parallel lines have lengths a and c, with $c>a$, and the equal legs have length b. The radius of the circumscribing circle is denoted by R.

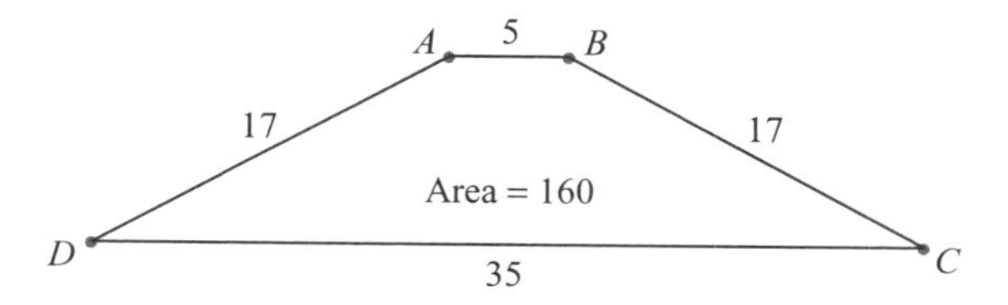

Fig. 6.3 An isosceles trapezium with integer sides and integer area.

Writing the distance between the two parallel lines in two different ways, we have

$$\sqrt{R^2 - \frac{1}{4}a^2} \pm \sqrt{R^2 - \frac{1}{4}c^2} = \sqrt{b^2 - \frac{1}{4}(c-a)^2}\,. \tag{6.2.9}$$

The plus sign is when the two parallel sides have a parallel diameter between them and the minus sign is when the parallel lines are both on the same side of a parallel diameter. Squaring both sides, simplifying, and squaring again, we find that

$$R^2 = \frac{b^2(b^2 + ac)}{4b^2 - (c-a)^2}\,. \tag{6.2.10}$$

We can make R rational by choosing integers x and y such that

$$4b^2 = y^2 + (c-a)^2 \tag{6.2.11}$$

and

$$4x^2 = y^2 + (c+a)^2\,, \tag{6.2.12}$$

for then $ac = x^2 - b^2$ and $R = bx/y$.

Working modulo 4, we see that all of $c - a$, $c + a$, and y must be even, so putting $c - a = 2l$, $c + a = 2m$, and $y = 2n$ we find that $b^2 = l^2 + n^2$ and $x^2 = m^2 + n^2$. Observe now that the problem of making R rational is identical to the problem of manufacturing a Heron triangle: two different Pythagorean triples have to be matched with equal short legs. As this is covered in Section 1.3 we do not give the details again. Instead, we give an illustrative example. Let $n = 8$, $l = 6$, and $m = 15$; then $x = 17$ and $b = 10$. It follows that $c = 21$ and $a = 9$. We find that $R^2 = 100 \times 289/256$, so that $R = 85/8$. Since the value of R^2 has been obtained by squaring twice, the question of whether the plus sign or minus sign holds in eqn (6.2.9) is one that has to be settled afterwards by substituting back. In the numerical example given above the minus sign holds, since $9.625 - 1.625 = 8$. From the preceding paragraph we see that

$$[ABCD] = \frac{1}{2}(a+c)\sqrt{b^2 - \frac{1}{4}(c-a)^2} = mn = 120\,.$$

Enlarging the figure by a factor of eight gives a figure with integer circumradius. By Ptolemy's theorem, x is the diagonal of the trapezium, which is thus also an integer. This example is illustrated in Fig. 6.4.

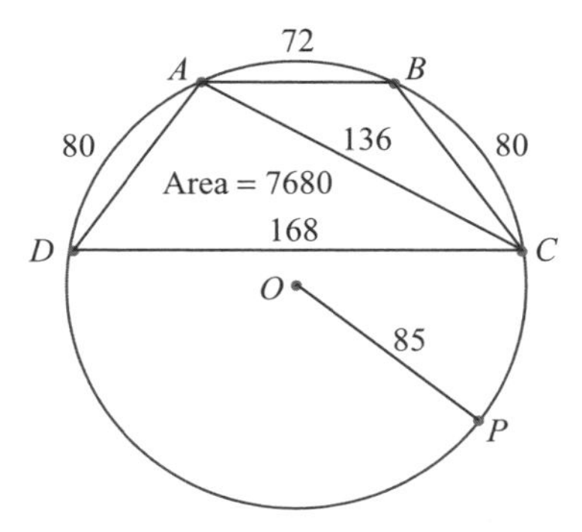

Fig. 6.4 An isosceles trapezium with integer sides, integer diagonals, integer area, and integer circumradius.

We have succeeded in showing that cyclic quadrilaterals, other than rectangles, exist in which the sides, the diagonals, the area, and the radius of the circumscribing circle are all integers.

Exercises 6.2

6.2.1 Use the parameters $r = 5$, $s = 4$, $t = 8$, $u = 10$, and all others equal to 1 in eqns (6.2.5) to find a cyclic quadrilateral with integer sides and integer area.

6.2.2 Match up the Pythagorean triples $(5, 12, 13)$ and $(12, 35, 37)$ to find an isosceles trapezium with integer sides, integer area, integer diagonals, and integer circumradius.

6.3 Equal sums of squares on the sides of a triangle

In this section we consider two problems associated with an integer-sided triangle ABC, and a point P in the plane of the triangle (not on its sides and not necessarily interior to the triangle). Let the feet of the perpendiculars from P onto the sides BC, CA, and AB be denoted by L, M, and N, respectively. We show how to determine the side lengths of ABC and the position of the point P such that the lengths of all of the line segments BL, LC, CM, MA, AN, and NB are integers. We also show how to locate P when these six line segments have given integer length.

In the configuration just described the following theorem holds.

Theorem 6.3.1 *Let L, M, and N be points on the sides BC, CA, and AB, respectively, of the triangle ABC. Then L, M, and N are the feet of the perpendiculars from a point P onto the sides of the triangle if and only if*

$$BL^2 + CM^2 + AN^2 = LC^2 + MA^2 + NB^2 . \tag{6.3.1}$$

□

The proof of the necessity of the condition depends only on multiple applications of Pythagoras' theorem. For the sufficiency, let the perpendiculars from L and M meet at P; then drop the perpendicular from P onto AB to meet it at N'; then use the given relation and the similar necessary relation with N' replacing N to show that N and N' coincide. A full proof is given in Durell (1946).

If the areal co-ordinates of P are (l, m, n), normalised so that $l + m + n = 1$, then it may be shown, see Example A.9.2 in the appendix, that the lengths in the theorem are given by the following expressions:

$$\begin{aligned} BL &= an + \frac{l}{2a}(c^2 + a^2 - b^2), & CM &= bl + \frac{m}{2b}(a^2 + b^2 - c^2), \\ AN &= cm + \frac{n}{2c}(b^2 + c^2 - a^2), & LC &= am + \frac{l}{2a}(a^2 + b^2 - c^2), \\ MA &= bn + \frac{m}{2b}(b^2 + c^2 - a^2), & NB &= cl + \frac{n}{2c}(c^2 + a^2 - b^2). \end{aligned} \tag{6.3.2}$$

From eqns (6.3.2) we see that if a, b, and c are integers and if l, m, and n are rational then the lengths of the six line segments are rational. The sides of the triangle may then be enlarged by a positive integral scale factor so that the line segments all have integer length. As an example, if $a = 96$, $b = 144$, $c = 192$, $l = 1/2$, $m = 1/3$, and $n = 1/6$ then $BL = 82$, $CM = 64$, $AN = 85$, $LC = 14$, $MA = 80$, and $NB = 107$, see Fig. 6.5.

The parametric solution of the equation $BL^2 + CM^2 + AN^2 = LC^2 + MA^2 + NB^2$ is given in terms of seven integer parameters by

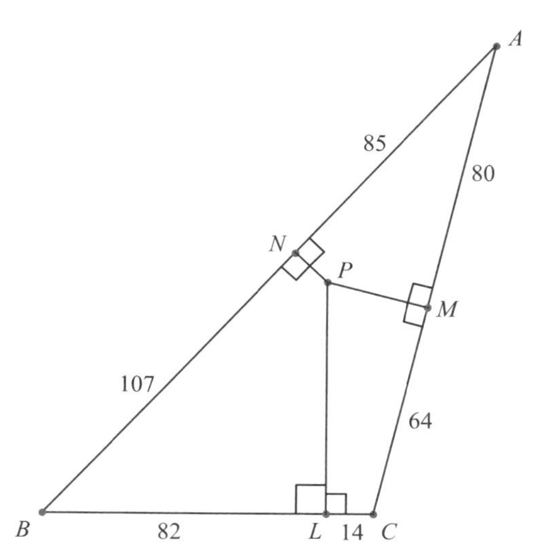

Fig. 6.5 Equal sums of squares on the sides of a triangle: $82^2 + 64^2 + 85^2 = 14^2 + 80^2 + 107^2$.

$$
\begin{aligned}
BL &= mf + vg - uh\,,\\
CM &= -vf + mg + th\,,\\
AN &= uf - tg + mh\,,\\
LC &= mf - vg + uh\,,\\
MA &= vf + mg - th\,,\\
NB &= -uf + tg + mh\,.
\end{aligned}
\tag{6.3.3}
$$

For a derivation of this parameter system see Bradley (1998). For example, when $m = 5$, $f = 1$, $g = 3$, $h = 3$, $t = 2$, $u = 2$, and $v = 3$ we obtain $8^2 + 18^2 + 11^2 = 2^2 + 12^2 + 19^2$. Given a set of integer values satisfying eqn (6.3.1), it is possible to locate the co-ordinates of the point P where the perpendiculars through L, M, and N to BC, CA, and AB, respectively, meet. In the example with $BL = 8$, $LC = 19$, $CM = 18$, $MA = 2$, $AN = 11$, and $NB = 12$ we have $a = 27$, $b = 20$, and $c = 23$, and using eqns (6.3.2) we find that $l = 603/1120$, $m = 27/56$, and $n = -23/1120$. Since l, m, and n arise as the solution of linear equations over the rational field, they must be rational. The negative value for n means that the perpendiculars meet outside the triangle on the other side of AB from C.

Exercise 6.3

6.3.1 Given $BL = 12$, $LC = 2$, $CM = 7$, $MA = 9$, $AN = 6$, and $NB = 12$, find the areal co-ordinates of the point P where the perpendiculars through L, M, and N to BC, CA, and AB, respectively, meet.

6.4 The integer-sided equilateral triangle

Two problems are now considered, which turn out to be equivalent algebraically, so that the solution to one problem provides a solution to the other.

The first problem is to find integer-sided equilateral triangles ABC and points P, not at the vertices, such that AP, BP, and CP are all integers. This is the same problem (by reducing the figure) as finding points P in the plane of an equilateral triangle of unit side, not at the vertices, that are at a rational distance from all three vertices. Solutions are easy to find when P lies on a side of the triangle, or when P lies on the circumcircle of ABC. The general problem of finding such points turns out to be considerably more difficult, but we are able to show that there are an infinite number of points P internal to the triangle with the required property.

This general problem of the internal point turns out to be equivalent algebraically to the problem of finding integer-sided triangles ABC for which the distances AF, BF, and CF from the vertices to the Fermat point F are all integers. For a triangle with angles all less than $120°$, the *Fermat* (or *Steiner*) *point* F is the point P internal to ABC which minimises the sum $AP + BP + CP$. The point F has the property that $\angle BFC = \angle CFA = \angle AFB = 120°$.

The equilateral triangle

Let ABC be an equilateral triangle of integer side d and suppose that the point P has (unnormalised) areal co-ordinates (l, m, n). Then we have

$$AP^2 = \frac{d^2(m^2 + mn + n^2)}{(l + m + n)^2}, \tag{6.4.1}$$

with similar expressions by cyclic change of l, m, and n for BP^2 and CP^2. See Section A.8 of the appendix for details of how to obtain these formulae. The problem of finding a single triangle ABC and a single point P is therefore solved if we can find integers l, m, and n (with not more than one of them zero) so that all of m^2+mn+n^2, n^2+nl+l^2, and l^2+lm+m^2 are perfect squares. We can then choose the side length $d = l + m + n$ to make each of AP, BP, and CP an integer.

The general problem of finding an infinite number of internal points P, that are a rational distance from the vertices of an equilateral triangle of unit side, depends on finding an infinite number of sets of *positive* integers l, m, and n such that all of $m^2 + mn + n^2$, $n^2 + nl + l^2$, and $l^2 + lm + m^2$ are perfect squares. The required points P then have areal co-ordinates $(l, m, n)/(l + m + n)$. This is because the normalised areal co-ordinates of a point P internal to ABC satisfy $0 < l, m, n < 1$ and $l + m + n = 1$.

Points on the sides of the triangle

One solution, which may be thought of as the trivial solution, is when P lies on a side of the equilateral triangle. For example, if P lies on BC and has areal co-ordinates $(0, 5/8, 3/8)$, then $BP/PC = 3/5$ and, from eqn (6.4.1), $AP = 7d/8$. Hence, if we take $d = 8$ then $AP = 7$, $BP = 3$, and $CP = 5$, see Fig. 6.6. Observe that AP, BP, and CP form the sides of an integer-sided triangle with the angle between BP and CP equal to $120°$. Such integer-sided triangles are documented in Section 1.2. Every solution to that problem may be used to create points on each side of an equilateral triangle with the required properties. Hence (reducing the size of the triangles by a scale factor of d) we have the following theorem.

Theorem 6.4.1 *Given an equilateral triangle ABC of unit side, there exist an infinity of points P on each side of the triangle such that all of AP, BP, and CP are rational. For each such point P on the side BC the triangle with sides AP, BP, and CP has an angle of $120°$ between the segments BP and CP.* □

Points on the circumcircle of the triangle

We consider a point P on the minor arc BC so that $ABPC$ is a cyclic quadrilateral. Then $\angle BPC = 120°$, since this angle supplements $\angle BAC$. Furthermore, $\angle BPA = \angle BCA = 60°$ and similarly $\angle CPA = 60°$. By Ptolemy's theorem, $AP = BP+CP$. So when BP and CP are made rational it follows immediately that AP is rational. In

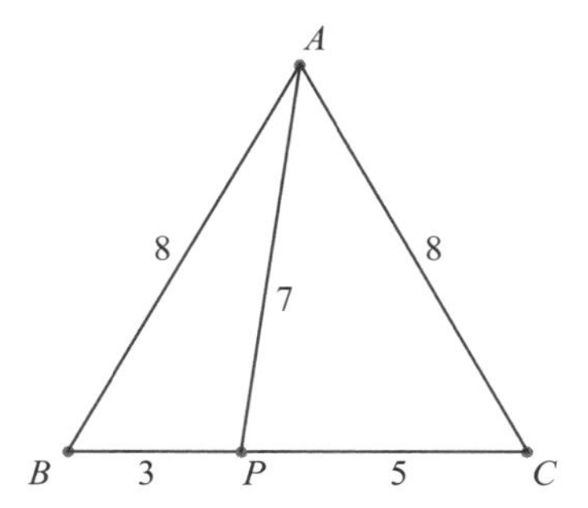

Fig. 6.6 An equilateral triangle with integer side and a point P on BC with AP, BP, and CP integers.

fact, a study of Table 1.2 shows that, for each integer-sided triangle with an angle of 120°, there exist two integer-sided triangles with an angle of 60° sharing a common shorter side and a common longer side.

For example, the integer-sided triangle having an angle of 120° with sides 3, 5, and 7 is complemented by two integer-sided triangles having an angle of 60°, namely the $(3, 8, 7)$ and the $(5, 8, 7)$ triangles. It follows, in this example, that if we choose $d = 7$ then a point P lies on the minor arc BC such that $BP = 3$, $CP = 5$, and $AP = 8$, see Fig. 6.7. Also, for every integer-sided triangle having an angle of 120°, there is a corresponding point P.

Hence (reducing the size of the triangles by a scale factor of d) we have the following theorem.

Theorem 6.4.2 *Given an equilateral triangle ABC of unit side, there exist an infinity of points P on each of the minor arcs BC, CA, and AB of the circumcircle such that all of AP, BP, and CP are rational.* □

Another way of looking at this result is as follows. Whenever P lies on the

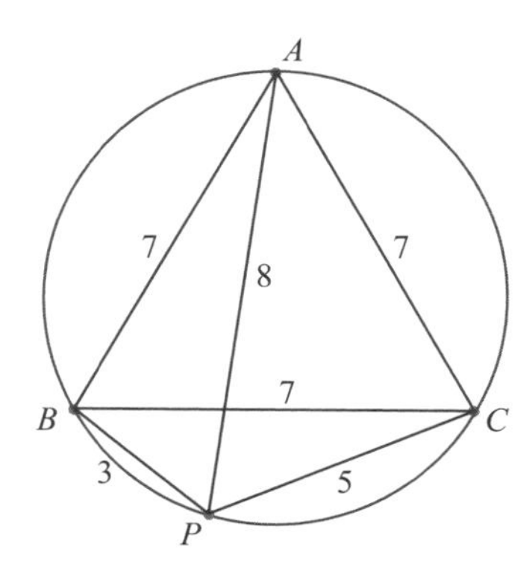

Fig. 6.7 An equilateral triangle with integer side and a point P on the circumcircle with AP, BP, and CP integers.

minor arc BC and so has its first areal co-ordinate negative (see Section A.4 of the appendix), there is a corresponding set of positive integers l, m, and n such that all of $m^2 + mn + n^2$, $n^2 - nl + l^2$, and $l^2 - lm + m^2$ are simultaneously perfect squares.

As pointed out earlier, it is an interesting question as to whether it is possible to find strictly positive integers l, m, and n, so that all of $m^2 + mn + n^2$, $n^2 + nl + l^2$, and $l^2 + lm + m^2$ are simultaneously perfect squares. For each set l, m, and n there exists a corresponding point P internal to an equilateral triangle ABC of unit side such that all of AP, BP, and CP are rational. Later in this section we prove that there are an infinite number of such points.

The Fermat point

As stated earlier, the same question arises in another geometrical context. Is it possible to find an integer-sided triangle ABC with angles less than $120°$ such that, if F is the Fermat point of ABC, then AF, BF, and CF are all of integer length? We now find a condition on the sides a, b, and c for a solution to exist.

Let $AF = l$, $BF = m$, and $CF = n$. Then, since the Fermat point F has the property that $\angle BFC = \angle CFA = \angle AFB = 120°$, we have

$$m^2 + mn + n^2 = a^2\,, \quad n^2 + nl + l^2 = b^2\,, \quad l^2 + lm + m^2 = c^2\,. \qquad (6.4.2)$$

From the first two equations we have

$$(m - l)(l + m + n) = a^2 - b^2\,. \qquad (6.4.3)$$

Writing $d = l+m+n$ we have $m-l = (a^2-b^2)/d$, and similarly $n-m = (b^2-c^2)/d$ and $l - n = (c^2 - a^2)/d$. It follows that $3l = d + (b^2 + c^2 - 2a^2)/d$, with similar equations for $3m$ and $3n$ by cyclic change of a, b, and c. Substituting for m and n in the first of eqns (6.4.2) we find, after some algebra, that

$$d^4 - (a^2 + b^2 + c^2)d^2 + (a^4 + b^4 + c^4 - b^2c^2 - c^2a^2 - a^2b^2) = 0\,. \qquad (6.4.4)$$

Solving this for d^2 we obtain

$$d^2 = \frac{1}{2}\left\{(a^2 + b^2 + c^2) + 4\sqrt{3}\,[ABC]\right\}, \qquad (6.4.5)$$

where we have used Heron's formula for the area $[ABC]$.

The condition on a, b, and c for rational l, m, and n to exist resolves into the single condition that the right-hand side of eqn (6.4.5) should be a perfect square. Solving the three equations (6.4.2) is thus reduced to finding integer solutions d of eqn (6.4.5). A computer search by J. T. Bradley (private communication) for solutions of eqn (6.4.5) with $1000 \geqslant a \geqslant b \geqslant c$ has revealed a number of solutions. The results are shown in Table 6.1. Since the two problems being treated are in correspondence, the tabulated values serve two purposes.

Firstly, they provide equilateral triangles of side d with a point P internal to triangle ABC such that $AP = a$, $BP = b$, and $CP = c$. Note that the smallest side length for

Table 6.1 Equilateral triangles with integer side d and a point P at integer distance from its vertices.

$AP = a$	$BP = b$	$CP = c$	Side d
73	65	57	112
95	88	73	147
152	147	43	185
205	168	127	283
208	185	97	273
280	221	111	331
296	285	49	331
343	312	95	403
361	315	296	559
387	343	152	485
407	392	323	645
437	377	147	520
469	464	285	691
473	343	255	592
485	408	247	637

such an equilateral triangle is 112. Secondly, they provide triangles of sides $BC = a$, $CA = b$, and $AB = c$ such that the Fermat distances AF, BF, and CF are rational and $d = AF + BF + CF$ is an integer.

A separate search to provide the least positive integer values of l, m, and n such that all of $m^2 + mn + n^2$, $n^2 + nl + l^2$, and $l^2 + lm + m^2$ are perfect squares has also been carried out and reveals eight solutions with $0 < l < m < n \leqslant 5016$. The results are shown in Table 6.2. This table provides integer-sided triangles with sides a, b, and c which possess integer Fermat distances $AF = l$, $BF = m$, and $CF = n$. It also provides equilateral triangles of side d in which there is an internal point P with $AP = a$, $BP = b$, and $CP = c$. Tables 6.1 and 6.2 are not independent, since the problems they solve are in correspondence. However, Table 6.2 gives solutions with

Table 6.2 Integer-sided triangles with integer Fermat distances AF, BF, and CF.

$AF = l$	$BF = m$	$CF = n$	a	b	c	d
195	264	325	511	455	399	784
264	325	440	665	616	511	1029
384	805	1520	2045	1744	1051	2709
455	1824	2145	3441	2405	2089	4424
360	1015	3864	4459	4056	1235	5239
435	1656	4669	5681	4901	1911	6760
1272	2065	4928	6223	5672	2917	8265
765	1064	5016	5624	5439	1591	6845

least integer values of l, m, and n, and Table 6.1 gives solutions with least integer values of a, b, and c (which correspond to rational l, m, and n that are not integers).

In Table 6.2, $d = AP + BP + CP$ is a minimum when P is at F. This is the minimum as the point P varies inside a triangle ABC with sides a, b, and c and with angles less than $120°$. It provides cases in which the distances AF, BF, and CF are integral. In both Tables 6.1 and 6.2 an equilateral triangle of side d has an internal point P such that $AP = a$, $BP = b$, and $CP = c$. Illustrative examples are provided by Figs 6.8 and 6.9.

The question arises as to whether there are an infinite number of points internal to an equilateral triangle of unit side that are a rational distance from all three vertices. The problem is similar to that of the rectangular box with integer sides and integer face diagonals, and to that of the triangle with integer sides and integer medians. I am indebted to Kevin Buzzard (private communication) for an analysis of the problem. In the language of number theory, what one has is a K3 surface, and the first observation is that there are lots of degenerate solutions. For example, if one puts $m = 0$, then

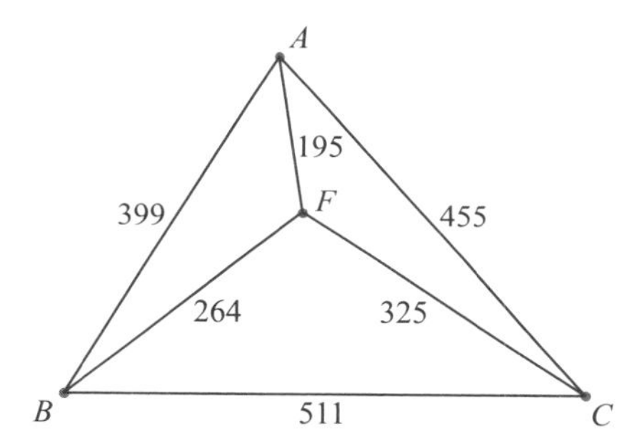

Fig. 6.8 A triangle with integer sides and integer Fermat distances.

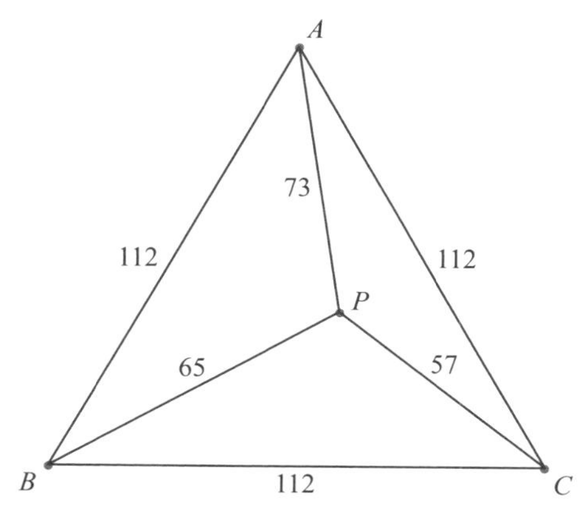

Fig. 6.9 An equilateral triangle with integer side and an internal point P with AP, BP, and CP integers.

$$\begin{aligned} l &= \frac{2s+1}{s^2+s+1}, \\ n &= \frac{s^2-1}{s^2+s+1}, \end{aligned} \tag{6.4.6}$$

for any rational number s, gives rational values of l and n such that $n^2 + nl + l^2$ is a rational square. Such degenerate solutions correspond to points on the sides of the triangle that are a rational distance from all three vertices. If we now start from these values of l and n for fixed s then we must solve the equations

$$\begin{aligned} m^2 + ml + l^2 &= (m+e)^2, \\ m^2 + mn + n^2 &= (m+f)^2, \end{aligned} \tag{6.4.7}$$

for m, e, and f. Solving for m we obtain $m = (e^2 - l^2)/(l - 2e) = (f^2 - n^2)/(n - 2f)$. This is a cubic curve in the two variables e and f, and from the degenerate solution $e = \frac{1}{2}l$ and $f = \frac{1}{2}n$ (corresponding to $m = \infty$) one can obtain another by drawing the tangent line at this point to meet the cubic again. The result is

$$e = \frac{1}{2}l - \frac{l^2}{l+n}, \quad f = \frac{1}{2}n - \frac{n^2}{l+n}. \tag{6.4.8}$$

Substituting back, clearing denominators, and expressing l, m, and n in terms of s, we obtain the parametric solution

$$\begin{aligned} l &= 8s(s+2)(2s+1), \\ m &= (s^2+6s+2)(2-2s-3s^2), \\ n &= 8s(s-1)(s+1)(s+2). \end{aligned} \tag{6.4.9}$$

Unfortunately, this parametric solution never has l, m, and n all positive, so it corresponds to points in the exterior of the triangle that are rational distances from the vertices, showing that there are an infinite number of such points. However, if one performs the trick again, drawing the tangent line from the point given by eqn (6.4.9) then one obtains the following parametric solution (after changing the signs of all of l, m, and n):

$$\begin{aligned} l &= 336s^{11} + 2760s^{10} + 10\,768s^9 + 23\,936s^8 + 30\,976s^7 + 24\,256s^6 \\ &\qquad + 13\,952s^5 + 8704s^4 + 4864s^3 + 1664s^2 + 256s \\ &= 8s(s+2)(2s+1)(21s^8 + 120s^7 + 352s^6 + 496s^5 + 344s^4 \\ &\qquad + 160s^3 + 128s^2 + 64s + 16)\,, \\ m &= 27s^{12} + 156s^{11} - 256s^{10} - 1944s^9 - 1516s^8 + 5600s^7 + 12\,032s^6 \\ &\qquad + 8000s^5 + 656s^4 - 1600s^3 - 1024s^2 - 384s - 64 \\ &= (s^2 - 2s - 2)^2(s^2 + 6s + 2)(3s^2 + 2s - 2)(9s^4 + 28s^3 + 32s^2 + 8s + 4)\,, \\ n &= 168s^{12} + 1296s^{11} + 4568s^{10} + 8304s^9 + 5952s^8 - 2816s^7 - 7104s^6 \\ &\qquad - 4224s^5 - 2432s^4 - 2304s^3 - 1152s^2 - 256s \\ &= 8s(s-1)(s+1)(s+2)(21s^8 + 120s^7 + 352s^6 + 496s^5 + 344s^4 \\ &\qquad + 160s^3 + 128s^2 + 64s + 16)\,. \end{aligned} \tag{6.4.10}$$

The corresponding values of a, b, and c are

$$\begin{aligned} a &= 183s^{12} + 1380s^{11} + 4380s^{10} + 7200s^9 + 6076s^8 + 416s^7 - 4256s^6 \\ &\qquad - 2816s^5 + 1424s^4 + 3136s^3 + 1984s^2 + 512s + 64\,, \\ b &= 8s(21s^{11} + 183s^{10} + 775s^9 + 1954s^8 + 3128s^7 + 3384s^6 + 2632s^5 \\ &\qquad + 1616s^4 + 912s^3 + 496s^2 + 176s + 32)\,, \\ c &= 27s^{12} + 324s^{11} + 2692s^{10} + 10\,384s^9 + 22\,620s^8 + 31\,904s^7 + 31\,136s^6 \\ &\qquad + 21\,120s^5 + 9680s^4 + 2624s^3 + 576s^2 + 256s + 64\,. \end{aligned} \tag{6.4.11}$$

Observe that, for sufficiently large values of s, all of l, m, and n are positive. We have therefore proved the following theorem.

Theorem 6.4.3 *Given an equilateral triangle ABC with unit side, there exist an infinite number of points P internal to ABC such that all of AP, BP, and CP are rational.* □

The following theorem, which we state without proof, is also true.

Theorem 6.4.4 *For each point P satisfying Theorem 6.4.3 there are three lines through P each passing through a vertex and each containing an infinite number of points Q such that all of AQ, BQ, and CQ are rational.* □

There is, of course, no intention to suggest that the above parameterisation is either the best or most efficient possible. It is not the numbers, but the idea that counts.

Exercises 6.4

6.4.1 Find strictly positive integers l, m, and n (not in the above tables) such that $l^2 + lm + m^2$ and $l^2 + ln + n^2$ are both perfect squares, and interpret the result geometrically.

6.4.2 Find the side of an equilateral triangle if a point P lies within it and $AP = 42$, $BP = 42$, and $CP = 12$.

7 Touching circles and spheres

It has been known for a long time, see Pedoe (1970), that the curvatures of four mutually touching circles are related by a single straightforward equation. The result was first obtained by Descartes (1901). In this chapter we show that infinite chains of touching circles, all with integer curvatures, can be formed. This means that the circles involved all have rational radii. Furthermore, we show how to obtain all such chains.

In the first section we deal with the singular case when one of the circles degenerates into a line. In the next section we deal with the case of sequences of mutually touching circles in two dimensions. We then go on to show how to generalise to sequences of mutually touching spheres in three dimensions, and finally to mutually touching hyperspheres in four dimensions. Infinite chains exist in all of these cases, and recurrence relations are established to show how they are formed.

The final section of this chapter employs the idea of the curvature of mutually touching circles to develop a three-parameter system for the sides of any Heron triangle up to similarity. The sides a, b, and c are obtained as homogeneous polynomials of the fourth degree in terms of the three integer parameters. In these expressions, common factors may have to be removed or enlargement factors introduced, which accounts for why the method only gives all Heron triangles up to similarity.

7.1 Three circles touching each other and all touching a line

We first consider a special case, when one of four circles in a plane degenerates into a line, so that we have a configuration of three circles which touch each other and which all touch a given line. The situation is a straightforward one, and the algebra involved is sometimes used in schools as an advanced exercise on Pythagoras' theorem. We are interested in cases in which the radii of the circles are rational numbers. Since the *curvature* of a circle is defined as the reciprocal of its radius, we can restrict consideration to curvatures that are positive integers. (When the curvatures of a finite number of circles are rational, the configuration can be reduced in size by a rational scale factor so that the curvatures of all the circles become integers.)

Suppose that the circles have radii a, b, and c, with $a \geqslant b > c$, and that their centres are respectively at the points with co-ordinates (h, a), (k, b), and $(0, c)$, so that the line that touches all the circles has equation $y = 0$. Since the circles touch, the three equations relating the squares of the distances between the centres of the three circles are

$$\begin{aligned} h^2 + (a-c)^2 &= (a+c)^2\,, \\ k^2 + (b-c)^2 &= (b+c)^2\,, \\ (h-k)^2 + (a-b)^2 &= (a+b)^2\,. \end{aligned} \tag{7.1.1}$$

Hence

$$h^2 = 4ac\,, \quad k^2 = 4bc\,, \quad h^2 - 2hk + k^2 = 4ab\,. \tag{7.1.2}$$

It follows that $hk = 2(ac + bc - ab)$, and squaring and eliminating h^2k^2 we obtain $(ab - ac - bc)^2 = 4abc^2$. On multiplying out and simplifying, we obtain

$$\frac{1}{a^2} + \frac{1}{b^2} + \frac{1}{c^2} = \frac{2}{bc} + \frac{2}{ca} + \frac{2}{ab}\,. \tag{7.1.3}$$

Hence, if we define the curvatures $\kappa = 1/a$, $\lambda = 1/b$, and $\mu = 1/c$, then we have

$$\kappa^2 + \lambda^2 + \mu^2 = 2(\lambda\mu + \mu\kappa + \kappa\lambda)\,, \tag{7.1.4}$$

where $\mu > \lambda \geqslant \kappa$. Since eqn (7.1.4) is homogeneous, we can choose κ, λ, and μ to be integers, as multiplying each of them by the same amount simply produces a similar figure to the first. Solving for μ we obtain $\mu = (\sqrt{\lambda} + \sqrt{\kappa})^2$, where the plus sign is chosen since the circle of radius c is assumed to lie between the circles of radii a and b, and hence has a larger curvature. In order to obtain rational radii for all three circles we must choose $\mu = z^2$, $\lambda = y^2$, and $\kappa = x^2$, where x, y, and z are positive integers, and then $z = x + y$. Transferring back to the original variables, we find that $a = 1/x^2$, $b = 1/y^2$, and $c = 1/z^2$ ($z > y \geqslant x$). Also, the positions of the centres are $(-2/zx, 1/x^2)$, $(2/zy, 1/y^2)$, and $(0, 1/z^2)$. So, for example, if we start with two circles of radius 1 and put a circle between them then its radius is $1/4$. If we then put

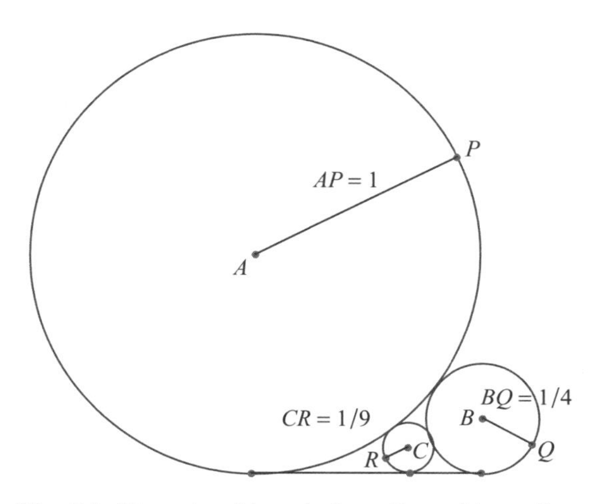

Fig. 7.1 Three touching circles, all touching a line.

a circle between the first and third circles its radius is $1/9$, and so on. In this way we obtain a chain of touching circles, all with rational radii, see Fig. 7.1.

Exercise 7.1

7.1.1 Starting with two circles C_1 and C_2 of radii 1 and $1/4$, respectively, touching each other externally and both touching a line, insert circles C_n, $n \geqslant 3$, between C_{n-1} and C_{n-2}, touching them both externally and touching the same line. Find the radii of C_3 and C_4, and find a formula for the radius of C_n.

7.2 Four circles touching one another externally

We now consider the general case in which there are four mutually touching circles in a plane. In this section we show how to find configurations in which all four circles have integer radius. The analysis is presented for the case in which all of the circles touch each other externally. However, we point out what happens when the fourth circle has internal contact with the other three, or when there are two circles that touch the other three externally.

Suppose that the three outside circles have radii a, b, and c and centres A, B, and C, respectively. Let the circle that touches them externally and that lies in the space between them have radius r and centre X. Then $AB = a + b$, $BC = b + c$, $CA = c + a$, $XA = a + r$, $XB = b + r$, and $XC = c + r$. (Note that in this section a, b, and c are **not** the sides of the triangle ABC.) Using Heron's formula (1.3.1) for the area of a triangle, we have $[ABC] = \sqrt{abc(a + b + c)}$, with similar expressions for $[XBC]$, $[XCA]$, and $[XAB]$. Equating the areas, it follows that

$$\sqrt{bcr(b + c + r)} + \sqrt{car(c + a + r)} + \sqrt{abr(a + b + r)} = \sqrt{abc(a + b + c)}\,. \tag{7.2.1}$$

If we make the substitutions $a = 1/\alpha$, $b = 1/\beta$, $c = 1/\gamma$, and $r = 1/\rho$, so that α, β, γ, and ρ are the curvatures of the four circles, then eqn (7.2.1) becomes

$$\alpha\sqrt{\beta\gamma + \gamma\rho + \rho\beta} + \beta\sqrt{\gamma\alpha + \alpha\rho + \rho\gamma} + \gamma\sqrt{\alpha\beta + \beta\rho + \rho\alpha} = \rho\sqrt{\alpha\beta + \beta\gamma + \gamma\alpha}\,. \tag{7.2.2}$$

If we can find solutions to eqn (7.2.2) in which all of α, β, γ, and ρ are integers, then by enlargement we obtain solutions of eqn (7.2.1) in which all of a, b, c, and r are integers; conversely, if the circles have integer radii then a similar figure exists in which the curvatures are all integers.

Suppose now that α, β, and γ are integers and

$$\sqrt{\alpha\beta + \beta\gamma + \gamma\alpha} = u\,. \tag{7.2.3}$$

Then we claim that $\rho = \alpha + \beta + \gamma + 2u$. In fact, given this value of ρ, we have

$$\begin{aligned}\beta\gamma+\gamma\rho+\rho\beta &= (\beta+\gamma)(\alpha+\beta+\gamma+2u)+\beta\gamma\\ &= \beta^2+2\beta\gamma+\gamma^2+\alpha\beta+\beta\gamma+\gamma\alpha+2u(\beta+\gamma)\\ &= (u+\beta+\gamma)^2\,.\end{aligned}$$

Hence $\sqrt{\beta\gamma+\gamma\rho+\rho\beta}=u+\beta+\gamma$, and similarly for the other square roots. The left-hand side of eqn (7.2.2) is now

$$\alpha(u+\beta+\gamma)+\beta(u+\gamma+\alpha)+\gamma(u+\alpha+\beta)=(\alpha+\beta+\gamma)u+2(\alpha\beta+\beta\gamma+\gamma\alpha)$$

and the right-hand side of eqn (7.2.2) is $u(\alpha+\beta+\gamma+2u)=(\alpha+\beta+\gamma)u+2u^2$, which is equal to the left-hand side by virtue of eqn (7.2.3). Note that, with this value of ρ, we have $(\rho-\alpha-\beta-\gamma)^2=4u^2=4(\alpha\beta+\beta\gamma+\gamma\alpha)$, and hence

$$2(\alpha^2+\beta^2+\gamma^2+\rho^2)=(\alpha+\beta+\gamma+\rho)^2\,. \qquad (7.2.4)$$

Furthermore, ρ is an integer if and only if u is an integer, so we definitely require $\alpha\beta+\beta\gamma+\gamma\alpha$ to be a perfect square for all of the curvatures to be integers.

Equation (7.2.4), due to Descartes (1901), is a quadratic in each variable, so it allows two possibilities for each variable given values of the others. This corresponds to the fact that, given any three touching circles, it is possible, in general, to draw two circles touching the other three. If the product of the two curvatures is positive then they touch externally, and if the product of the two curvatures is negative then they touch internally. See Pedoe (1970) for proofs of these remarks. Given positive values of α, β, and γ, the second value of ρ from eqn (7.2.4) is given by $\rho=\alpha+\beta+\gamma-2u$. If positive, it corresponds to the fact that there is a second circle of much larger radius touching the three given circles externally; if negative, it corresponds to the fact that the second circle surrounds the three given circles, thereby touching them internally. The second value of ρ obtained from Descartes' equation corresponds to eqn (7.2.2) only when some of the square roots are made negative. This corresponds to the fact that X now lies outside the triangle ABC. The larger value of ρ corresponds to the case which we are considering, since it corresponds to a circle of smaller radius than the three given circles, and lies in the space between them.

It is a strange fact that, if integers α, β, and γ are chosen so that $\alpha\beta+\beta\gamma+\gamma\alpha$ is a perfect square, then we can always find two partners $\rho=\alpha+\beta+\gamma\pm 2\sqrt{\alpha\beta+\beta\gamma+\gamma\alpha}$, one with the plus sign and one with the minus sign, so that all of $\alpha\beta+\beta\rho+\rho\alpha$, $\beta\gamma+\gamma\rho+\rho\beta$, and $\gamma\alpha+\alpha\rho+\rho\gamma$ are perfect squares. For example, with $\alpha=3$, $\beta=7$, and $\gamma=10$ we have $\rho=42$ or -2. With the choice $\rho=42$, the four squares are 121, 441, 784, and 576. To obtain integer radii we must multiply up by the least common multiple of their denominators. In geometrical terms, this means that, if we are given three circles touching externally of radii 70, 30, and 21, then the circle in the space between them and touching them has radius 5. The root $\rho=-2$ means that there is a large circle surrounding them of radius 105, touching the other three internally, see Fig. 7.2.

It is possible to continue the process, replacing one of the first three circles by the fourth and then obtaining another circle of integer curvature touching the second set

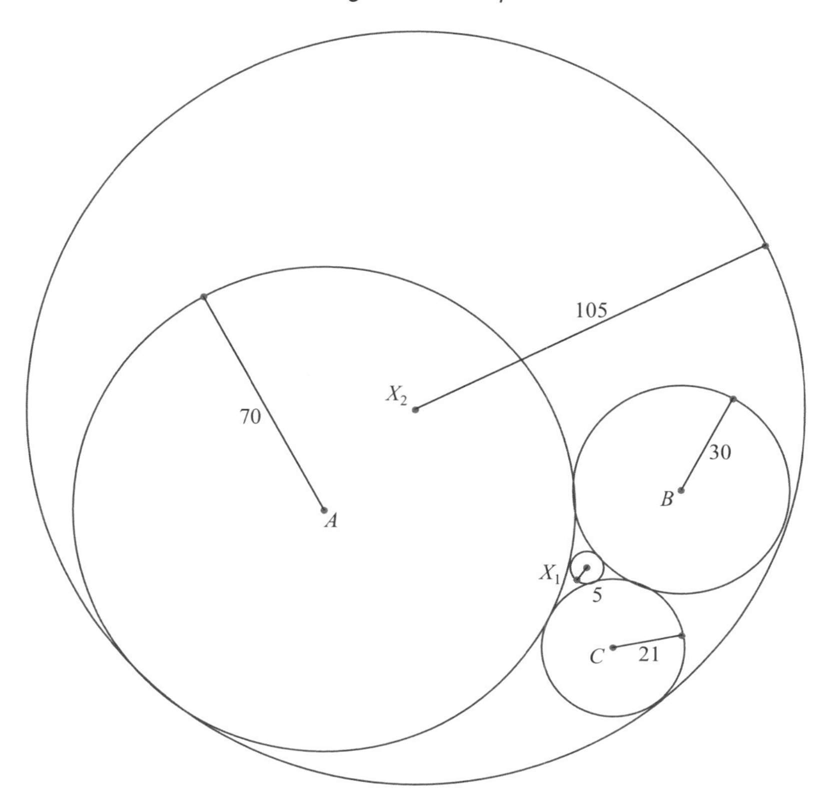

Fig. 7.2 Four touching circles.

of three circles. Continuing in this way we may obtain a chain of circles, and if there are a finite number of them then the configuration can be enlarged so that all of these circles have integer radius.

A parametric representation

It is possible to solve the equation $\alpha\beta + \beta\gamma + \gamma\alpha = u^2$ parametrically and hence to obtain sequences of four integer curvatures corresponding to four touching circles. It is shown how to obtain the parametric solution in Section 7.5, where the theory of this section is applied to a further study of Heron triangles. Taking the reciprocals of the curvatures and then multiplying up by the least common multiple of their denominators gives integer solutions for the radii in the problem of four touching circles.

The parametric solution is

$$
\begin{aligned}
\alpha &= f^2 + 4d^2 + 3ed + fd\,, \\
\beta &= e^2 + 4d^2 + 3fd + ed\,, \\
\gamma &= e^2 + f^2 + 2ef - 2d^2 - ed - fd\,, \\
u &= e^2 + f^2 + ef + 3ed + 3fd\,.
\end{aligned}
\tag{7.2.5}
$$

The two possible values of ρ are given by $4e^2 + 4f^2 + 4ef + 6d^2 + 9ed + 9fd$ and $6d^2 - 3ed - 3fd$.

For example, with $d = 3$, $e = -2$, and $f = -3$ we obtain $\alpha = 18$, $\beta = 7$, $\gamma = 22$, $u = 26$, and $\rho = 99$ or $\rho = -5$. Geometrically, this means that if we have three large circles of radii 63, 198, and 77 then the circle in the space between them and touching them has radius 14. By letting d, e, and f run through all sets of integers, we can obtain (by enlargement) all configurations in which four touching circles have integer radii.

Exercises 7.2

7.2.1 Given that $\alpha = 1$, $\beta = 4$, and $\gamma = 12$, find the two possible values of ρ, the variables having the meaning in the text. Interpret your result geometrically.

7.2.2 Put $e = 5$, $f = 4$, and $d = 1$ in the parametric representation (7.2.5) and find the corresponding values of α, β, γ, and ρ. Interpret your result geometrically.

7.3 Five spheres touching each other externally

We now consider the analogue in three dimensions of the theory in Section 7.2. What happens in three dimensions is that it is possible to find five mutually touching spheres, and we show how to obtain configurations in which these spheres all have integer radius. We start with a piece of algebra, which is later associated with the curvatures of the spheres.

For integers a, b, c, and d we define

$$
[a, b, c, d] = \sqrt{3(2ab + 2bc + 2ca + 2ad + 2bd + 2cd - a^2 - b^2 - c^2 - d^2)}\,.
\tag{7.3.1}
$$

Clearly, $[a, b, c, d]$ is either a non-negative integer, or irrational, or complex. It cannot be a rational unless it is an integer. Also, since $a^2 + b^2 + c^2 + d^2$ has the same parity as $a + b + c + d$, it follows that when $[a, b, c, d]$ is an integer it has the same parity as $a + b + c + d$, and hence $e = \frac{1}{2}(a + b + c + d + [a, b, c, d])$ is an integer also.

Five non-negative integers a, b, c, d, and e are said to form a *touching set* if

(i) $a \leqslant b \leqslant c \leqslant d \leqslant e$;

(ii) only a, if any, is zero;

(iii) $e = \frac{1}{2}(a + b + c + d + [a, b, c, d])$, where $[a, b, c, d]$ is given by eqn (7.3.1); and

(iv) $[a, b, c, d]$ is an integer.

Theorem 7.3.1 *If a, b, c, d, and e form a touching set, then all of $[b, c, d, e]$, $[a, c, d, e]$, $[a, b, d, e]$, and $[a, b, c, e]$, as well as $[a, b, c, d]$, are integers.*

Proof We already know, by definition, that $m = [a, b, c, d]$ is an integer. So, by symmetry, it is sufficient to show that $[b, c, d, e]$ is an integer. We have $[b, c, d, e]^2 = 6(bc + cd + db + be + ce + de) - 3(b^2 + c^2 + d^2 + e^2)$, which, on inserting the value of e and simplifying, is equal to

$$\begin{aligned}
&\frac{21}{2}(bc + cd + db) + \frac{3}{2}(ab + ac + ad) \\
&\quad + \frac{3m}{2}(b + c + d - a) - \frac{3}{4}(a^2 + b^2 + c^2 + d^2 + m^2) \\
&\qquad = \frac{1}{4}m^2 + \frac{3m}{2}(b + c + d - a) + \frac{9}{2}(bc + cd + db - ab - ac - ad) \\
&\qquad\qquad + \frac{9}{4}(a^2 + b^2 + c^2 + d^2) \\
&\qquad = \frac{1}{4}[m + 3(b + c + d - a)]^2 .
\end{aligned}$$

Hence $[b, c, d, e] = \frac{1}{2}(m + 3b + 3c + 3d - 3a)$, which is an integer. Similarly, $[a, c, d, e] = \frac{1}{2}(m + 3a + 3c + 3d - 3b)$, $[a, b, d, e] = \frac{1}{2}(m + 3a + 3b + 3d - 3c)$, and $[a, b, c, e] = \frac{1}{2}(m + 3a + 3b + 3c - 3d)$. □

Theorem 7.3.2 *Let a, b, c, d, and e be a touching set, as defined prior to Theorem 7.3.1. Then the sequence (u_n) defined by $u_1 = a$, $u_2 = b$, $u_3 = c$, $u_4 = d$, $u_5 = e$, and $u_{n+5} = u_{n+4} + u_{n+3} + u_{n+2} + u_{n+1} - u_n$, for all $n = 1, 2, 3, \ldots$, is such that u_n, u_{n+1}, u_{n+2}, u_{n+3}, and u_{n+4} is a touching set for all $n = 1, 2, 3, \ldots$.*

Proof The proof is by induction. The case $n = 1$ is given. As the induction hypothesis, suppose that u_n, u_{n+1}, u_{n+2}, u_{n+3}, and u_{n+4} is a touching set and that

$$u_{n+4} = \frac{1}{2}(u_n + u_{n+1} + u_{n+2} + u_{n+3} + v_n)\,, \tag{7.3.2}$$

where $v_n = [u_n, u_{n+1}, u_{n+2}, u_{n+3}]$. Since $[u_{n+1}, u_{n+2}, u_{n+3}, u_{n+4}] = v_{n+1}$ is an integer, we know from Theorem 7.3.1 that

$$2v_{n+1} = v_n + 3u_{n+3} + 3u_{n+2} + 3u_{n+1} - 3u_n\,. \tag{7.3.3}$$

Also, from the recurrence relation,

$$u_{n+5} = u_{n+4} + u_{n+3} + u_{n+2} + u_{n+1} - u_n\,. \tag{7.3.4}$$

Eliminating v_n from eqns (7.3.2) and (7.3.3), we obtain

$$v_{n+1} = u_{n+4} + u_{n+3} + u_{n+2} + u_{n+1} - 2u_n\,. \tag{7.3.5}$$

Eliminating u_n from eqns (7.3.4) and (7.3.5), we obtain

$$u_{n+5} = \frac{1}{2}(u_{n+1} + u_{n+2} + u_{n+3} + u_{n+4} + v_{n+1}), \tag{7.3.6}$$

and hence, by Theorem 7.3.1, the induction is complete. □

Now suppose that the radii of the five spheres that touch each other externally are k, l, m, n, and p, and define their curvatures by the equations $\kappa = 1/k$, $\lambda = 1/l$, $\mu = 1/m$, $\nu = 1/n$, and $\rho = 1/p$. Then the generalisation of Descartes' relation for the curvatures of the five touching spheres is

$$3(\kappa^2 + \lambda^2 + \mu^2 + \nu^2 + \rho^2) = (\kappa + \lambda + \mu + \nu + \rho)^2 . \tag{7.3.7}$$

Solving this for ρ we obtain a solution

$$\rho = \frac{1}{2}(\kappa + \lambda + \mu + \nu + [\kappa, \lambda, \mu, \nu]) . \tag{7.3.8}$$

When $\kappa \leqslant \lambda \leqslant \mu \leqslant \nu \leqslant \rho$ and all are positive integers, except possibly κ, which may be zero (corresponding to a plane), and $[\kappa, \lambda, \mu, \nu]$ is an integer, then κ, λ, μ, ν, and ρ form a touching set. For example, see Exercise 7.3.2, spheres of radii 6, 6, 6, 2, and 1 can be arranged to touch one another externally. Also, see Exercise 7.3.1, spheres of radii 12, 12, 4, and 3 can be arranged to touch each other externally, whilst all touching a plane. Given a set of five spheres touching each other externally, which have integer radii k, l, m, n, and p, then $lmnp$, $kmnp$, $klnp$, $klmp$, and $klmn$ must form a touching set. It follows that the touching sets are in correspondence with sets of five spheres touching each other externally and having integer radii.

There is a second value of ρ, which is a solution of eqn (7.3.7), given by

$$\rho = \frac{1}{2}(\kappa + \lambda + \mu + \nu - [\kappa, \lambda, \mu, \nu]) , \tag{7.3.9}$$

which, if negative, gives a fifth sphere having internal contact with the first four spheres.

A two-parameter system for a touching set, which can be used to start a sequence, as in Theorem 7.3.2, is given by

$$a = m^2 + mn , \quad b = n^2 + mn , \quad c = m^2 + mn + n^2 ,$$
$$d = 3m^2 + 5mn + 3n^2 , \quad e = 4m^2 + 7mn + 4n^2 ,$$

where m and n are non-negative integers. However, this is not a general solution.

Exercises 7.3

7.3.1 Show that 0, 1, 1, 3, and 4 is a touching set, and find the next three terms of the sequence defined in Theorem 7.3.2.

7.3.2 Show that 1, 1, 1, 3, and 6 is a touching set, and find the next five terms of the sequence defined in Theorem 7.3.2.

7.3.3 Given that 4, 9, 19, 25, and 57 forms a touching set, state the geometrical interpretation of this in terms of touching spheres.

7.3.4 Four touching spheres of radii 350, 175, 150, and 150 are all touched by a fifth sphere. Find possible values of its radius.

7.3.5 Prove that, if a, b, c, d, and e is a touching set, then

$$e[a, b, c, d] = a[b, c, d, e] + b[a, c, d, e] + c[a, b, d, e] + d[a, b, c, e]\,.$$

7.4 Six touching hyperspheres in four-dimensional space

We next investigate how the theory of the previous two sections generalises into four dimensions, when we expect to find six mutually touching hyperspheres and configurations in which their radii are all integers. This excursion into the abstract is prompted by the indicative properties of touching circles and spheres dealt with in the last two sections. The requirement that the curvatures should be integers led to four quadratic expressions of the form $lm+mn+nl$ being simultaneously perfect squares in the case of four touching circles. It also led to five quadratic expressions of the form $3(2ab+2bc+2ca+2ad+2bd+2cd-a^2-b^2-c^2-d^2)$ being simultaneously perfect squares in the case of five touching spheres. Furthermore, there is the curious property in each case that chains of such expressions can be formed. These have the geometrical interpretation that there are chains of circles and spheres, such that any four circles in a chain or any five spheres in a chain are mutually touching and by enlargement can be made to have integer radius. We show in this section how these results generalise further, though the curvatures sometimes turn out to be rational numbers that are not necessarily integers. This is no problem, however, as one finds that they can all be multiplied, at any given stage, by a factor of three to make them integral.

The equation to be satisfied by the curvatures of six mutually touching hyperspheres is

$$4(\kappa^2+\lambda^2+\mu^2+\nu^2+\rho^2+\sigma^2) = (\kappa+\lambda+\mu+\nu+\rho+\sigma)^2\,. \qquad (7.4.1)$$

We now define

$$[a, b, c, d, e] = \sqrt{2(ab + ac + ad + ae + bc + bd + be + cd + ce + de - a^2 - b^2 - c^2 - d^2 - e^2)}\,. \tag{7.4.2}$$

Then the solutions of eqn (7.4.1), given suitable values of λ, μ, ν, ρ, and σ, are

$$\kappa = \frac{1}{3}(\lambda + \mu + \nu + \rho + \sigma \pm 2[\lambda, \mu, \nu, \rho, \sigma])\,. \tag{7.4.3}$$

Observe that, when $[\lambda, \mu, \nu, \rho, \sigma]$ is an integer, κ may not be an integer but may be a rational with a denominator of 3. This does not matter as eqn (7.4.1) is homogeneous of degree two in the variables, so all of the curvatures may be multiplied by 3 afterwards to provide integer solutions.

We call non-negative integers a, b, c, d, e, and f a *touching set* if

(i) $0 \leqslant a \leqslant b \leqslant c \leqslant d \leqslant e \leqslant f$;

(ii) only a, if any, is zero;

(iii) $[a, b, c, d, e]$ is an integer, where $[a, b, c, d, e]$ is defined by eqn (7.4.2); and

(iv) $f = (1/3)(a + b + c + d + e + 2[a, b, c, d, e])$.

These integers correspond to the curvatures of six mutually touching hyperspheres in four-dimensional space, touching externally, with the hypersphere of smallest radius corresponding to the curvature f.

For example, if $a = 0$, $b = 1$, $c = 1$, $d = 1$, and $e = 2$, then $f = 3$. This corresponds to four mutually touching hyperspheres of radii 6, 6, 6, and 3 lying on a hyperplane, with a hypersphere of radius 2 touching the hyperplane and all of the other hyperspheres. As a second example, if $a = 1$, $b = 1$, $c = 1$, $d = 2$, and $e = 3$ then $f = 16/3$. We then produce a touching set by multiplying up by 3 to give $a = 3$, $b = 3$, $c = 3$, $d = 6$, $e = 9$, and $f = 16$. The geometrical interpretation of this is that hyperspheres of radii 48, 48, 48, 24, 12, and 9 can be arranged so that they touch each other. Analogues of Theorems 7.3.1 and 7.3.2 exist, and these hold for touching sets in four dimensions, as explained above.

Theorem 7.4.1 *If a, b, c, d, e, and f form a touching set, then all of $[b, c, d, e, f]$, $[a, c, d, e, f]$, $[a, b, d, e, f]$, $[a, b, c, e, f]$, and $[a, b, c, d, f]$, as well as $[a, b, c, d, e]$, are integers.* □

The proof is similar to that of Theorem 7.3.1 and is omitted.

Theorem 7.4.2 *If a, b, c, d, e, and f form a touching set and if we define the sequence (u_n) by $u_1 = a$, $u_2 = b$, $u_3 = c$, $u_4 = d$, $u_5 = e$, $u_6 = f$, and $u_{n+6} = (2/3)(u_{n+5} + u_{n+4} + u_{n+3} + u_{n+2} + u_{n+1}) - u_n$, $n = 1, 2, 3, \ldots$, then any six consecutive terms of the sequence, when multiplied, if necessary, by an appropriate power of 3, form a touching set.* □

Again, the proof is similar to that of Theorem 7.3.2 and is omitted.

Exercise 7.4

7.4.1 Show that 4, 12, 13, 45, 45, and 61 is a touching set for six hyperspheres in four dimensions. Find the next term in the sequence given by the recurrence relation defined in Theorem 7.4.2.

7.5 Heron triangles revisited

In Section 1.3 we defined a Heron triangle ABC to be a triangle with integer sides, integer area, and one in which the altitude from A is of integer length. **In this section we drop the last of these conditions and call a triangle a Heron triangle if it has integer sides and integer area**. With this definition, every Heron triangle is similar to one with the definition used in Section 1.3.

An application of the theory given in Section 7.2 enables us to derive formulae, in terms of four parameters, for all integer-sided triangles having integer area. One of these parameters is simply an enlargement factor, so we may regard the solution as one that depends essentially only on three parameters. The formulae for the side lengths are homogeneous expressions of degree four in terms of the parameters, or of degree five if you include the enlargement factor. In Section 1.3 it was shown how to construct Heron triangles that are the sum or difference of two integer-sided right-angled triangles. This gives geometrical insight into what Heron triangles are, but leads to increasingly complicated formulae for their side lengths, as a result of the different ways that the right-angled triangles can be fitted together. The theory of this section shows instead how Heron triangles arise from configurations of touching circles with integer radii, and because of this link it shows how chains of Heron triangles can be formed, provided that they are suitably enlarged.

Conversion of Heron's formula

We suppose that ABC is a Heron triangle. We start from Heron's formula

$$[ABC] = \frac{1}{4}\sqrt{(a+b+c)(b+c-a)(c+a-b)(a+b-c)}\,, \tag{7.5.1}$$

where a, b, and c are integers. Now put $a = m + n$, $b = n + l$, and $c = l + m$, where l, m, and n are the radii of three mutually touching circles, centred at A, B, and C, respectively. From Exercise 1.3.5, in any Heron triangle either a, b, and c are all even or two of a, b, and c are odd and one is even. Thus l, m, and n are positive integers. The change of variable produces the more manageable formula

$$[ABC] = \sqrt{lmn(l+m+n)}\,. \tag{7.5.2}$$

We now write $l = 1/u$, $m = 1/v$, and $n = 1/w$, where u, v, and w are the curvatures of the three touching circles. This now produces the formula

$$[ABC] = \frac{1}{uvw}\sqrt{vw + wu + uv}\,. \tag{7.5.3}$$

For example, if we start with the Heron triangle having $a = 14$, $b = 13$, and $c = 15$, then we obtain $l = 7$, $m = 8$, $n = 6$, $u = 1/7$, $v = 1/8$, $w = 1/6$, and $[ABC] = 84$. The formula in terms of u, v, and w is easy to handle, for, as we saw in Section 7.2, it is possible to find *integer* u, v, and w such that $\sqrt{vw + wu + uv}$ is an integer. However, u, v, and w are not integers, but are the reciprocals of integers. As an example shows, the matter is easily resolved. Suppose that we start with $u = 48$, $v = 42$, and $w = 56$ (by putting $u = mn$, $v = nl$, and $w = mn$); then $uv + vw + wu = 336^2$, a perfect square. This time $l = 1/48$, $m = 1/42$, and $n = 1/56$ are not integers. However, if we multiply them by the least common multiple of their denominators we get back to $l = 7$, $m = 8$, $n = 6$, $a = 14$, $b = 13$, and $c = 15$. It follows that Heron triangles are in 1–1 correspondence with positive integer solutions of the Diophantine equation

$$uv + vw + wu = x^2\,. \tag{7.5.4}$$

Solution in positive integers of the equation $uv + vw + wu = x^2$

In Section 7.2 we saw how to construct chains of solutions of eqn (7.5.4). If we start from any solution of $[u, v, w] = uv + vw + wu = x^2$ then $[v, w, u + v + w + 2x]$ is a perfect square.

In fact,

$$\begin{aligned}[v, w, u + v + w + 2x] &= vw + vu + v^2 + vw + 2vx + wu + vw + w^2 + 2wx \\ &= v^2 + 2vw + x^2 + 2vx + 2wx + w^2 \\ &= (v + w + x)^2\,.\end{aligned}$$

For example, $[-3, 4, 12] = 48 - 36 - 12 = 0$. So the next term in the sequence is $-3 + 4 + 12 + 0 = 13$. Then $[4, 12, 13] = 48 + 52 + 156 = 16^2$ and the next term in the sequence is $4 + 12 + 13 + 2 \times 16 = 61$ and $[12, 13, 61] = 41^2$.

One can also reverse the procedure. That is, if $[u, v, w] = x^2$, then

$$\begin{aligned}[u, v, u + v + w - 2x] &= uv + u^2 + uv + uw - 2ux + uv + v^2 + vw - 2vx \\ &= x^2 - 2ux - 2vx + u^2 + 2uv + v^2 \\ &= (u + v - x)^2\end{aligned}$$

is a perfect square. In the above example the chain started with $x = 0$, but not all chains of solutions of $[u, v, w] = x^2$ start with a trio satisfying $[u, v, w] = 0$. For example, $[-3, 5, 8] = 1$, but $u + v + w - 2x = 8$, so the chain based on this trio starts with $x = 1$ and not $x = 0$. Chains of solutions of eqn (7.5.4) are very intriguing, but they do not provide a general solution.

In Section 7.2 we gave a three-parameter solution of the equation $[u, v, w] = x^2$, without derivation. We now give the derivation. The method is similar to that used in

Section 1.2. It is to produce a non-singular linear change of variable that transforms the given equation to one that is linear in each variable. There are many possibilities. One such transformation is

$$\begin{aligned} u &= e + f + d\,, \\ v &= g + f - 2d\,, \\ w &= g + e - 2d\,, \\ x &= g + e + f\,. \end{aligned} \tag{7.5.5}$$

Substituting into $[u, v, w] = x^2$, we obtain $g = (5ed + 5fd - ef)/(e + f - 2d)$. If we now substitute this value for g into the expressions for u, v, w, and x and multiply up to clear denominators then we obtain

$$\begin{aligned} u &= e^2 + 2ef + f^2 - de - df - 2d^2\,, \\ v &= 3ed + fd + f^2 + 4d^2\,, \\ w &= 3fd + ed + e^2 + 4d^2\,, \\ x &= 3ed + 3fd + e^2 + f^2 + ef\,. \end{aligned} \tag{7.5.6}$$

With this parameterisation, u, v, w, and x may have a common factor that needs removing. Also, care must be taken not to choose parameters with $e + f = 2d$ or $-d$, for that would make $u = 0$, which is not allowed. For a non-degenerate triangle none of the variables is allowed to be zero, so other choices of parameters must be excluded. Also, from the geometry of the problem, the parameters must be chosen so that u, v, w, and x are positive integers.

Although the transformation (7.5.5) is not unimodular, this does not mean, by restricting d, e, and f to have integer values, that any solution is missing. This is because eqn (7.5.4) is homogeneous of degree 2, so when d, e, and f are rational with denominator 3 (the determinant of transformation (7.5.5)) they can each be multiplied by 3 and then all that happens is that u, v, w, and x have a common factor of 9. As no attempt has been made to provide parameters in which u, v, w, and x have no common factor and many of the solutions do have a common factor anyway, this is not an extra problem. Since eqn (7.5.4) is homogeneous, the solutions (7.5.6) are made complete by including a possible integer enlargement factor.

Examples using the parameters d, e, and f

Choose $d = 1$, $e = 2$, and $f = 3$. Then $u = 18$, $v = 22$, $w = 19$, and $x = 34$. This corresponds to $l = 1/18$, $m = 1/22$, and $n = 1/19$. We now multiply up by 3762 to get integral values $l = 209$, $m = 171$, and $n = 198$. From this we obtain a triangle with sides $a = m + n = 369$, $b = n + l = 407$, and $c = l + m = 380$. Finally, $[ABC] = 63\,954$.

As a second example, choose $d = 11$, $e = 83$, and $f = 41$. Then $u = 13\,370$, $v = 5355$, and $w = 9639$ and, cancelling a common factor of 153, we obtain $u = 90$, $v = 35$, and $w = 63$. Hence $l = 1/90$, $m = 1/35$, and $n = 1/63$. Multiplying up by

630, we obtain $l = 7$, $m = 18$, and $n = 10$. From this we obtain the triangle with sides $a = m + n = 28$, $b = n + l = 17$, and $c = l + m = 25$. Finally, $[ABC] = 210$. We see that this corresponds to the union of two right-angled triangles with sides 8, 15, and 17, and 20, 15, and 25. The second example is chosen deliberately to illustrate the fact that u, v, and w sometimes have a large common factor, so that relatively large values of d, e, and f may lead to triangles with short side lengths.

Parametric representation of the sides of a Heron triangle

Given u, v, and w for which $uv + vw + wu$ is a perfect square, we now know that a triangle with sides $a = u(v + w)$, $b = v(w + u)$, and $c = w(u + v)$ is a Heron triangle. With u, v, and w given in terms of d, e, and f, as defined above, we find a parametric representation for the sides of a Heron triangle.

This is

$$\begin{aligned}
a &= -16d^4 - 16d^3e - 16d^3f + 2d^2e^2 + 8d^2ef + 2d^2f^2 + 3de^3 \\
&\quad + 11de^2f + 11def^2 + 3df^3 + e^4 + 2e^3f + 2e^2f^2 + 2ef^3 + f^4 , \\
b &= 8d^4 + 6d^3e + 10d^3f + 8d^2e^2 + 14d^2ef + 8d^2f^2 + 6de^3 \\
&\quad + 8de^2f + 5def^2 + 3df^3 + 2e^2f^2 + 2ef^3 + f^4 , \\
c &= 8d^4 + 10d^3e + 6d^3f + 8d^2e^2 + 14d^2ef + 8d^2f^2 + 3de^3 \\
&\quad + 5de^2f + 8def^2 + 6df^3 + e^4 + 2e^3f + 2e^2f^2 .
\end{aligned} \tag{7.5.7}$$

Parameters must, of course, be chosen so that u, v, and w, and hence a, b, and c, are positive. The method ensures that the triangle inequalities are automatically satisfied.

With these values,

$$\begin{aligned}
[ABC] = |uvwx| &= |e^2 + 2ef + f^2 - de - df - 2d^2||3ed + fd + f^2 + 4d^2| \\
&\quad \times |3fd + ed + e^2 + 4d^2||3ed + 3fd + e^2 + f^2 + ef| .
\end{aligned} \tag{7.5.8}$$

The circumradius of triangle ABC is given by

$$R = \frac{(v + w)(w + u)(u + v)}{4x} . \tag{7.5.9}$$

In eqns (7.5.7) the values of a, b, and c may be multiplied by an additional enlargement factor k. Then the area is multiplied by k^2. As can be seen from one of the numerical examples above, although eqns (7.5.7) always give the sides of a Heron triangle, it may be one in which a common factor can be removed. In fact, the equations scarcely look as if a, b, and c have any low values whatsoever. In other words, these equations give all Heron triangles, but only up to similarity.

Exercises 7.5

7.5.1 Use the parameters $d = 2$, $e = 1$, and $f = 5$ in eqns (7.5.7) to find a Heron triangle and enlarge it so that an altitude is an integer. Hence show how the Heron triangle arises as the union of two integer-sided right-angled triangles with a common side.

7.5.2 Repeat Exercise 7.5.1 with $d = 1$, $e = 16$, and $f = 25$.

8 More on triangles

We now embark on an analysis of some more advanced properties of integer-sided triangles. Though there are a large number of problems that could be treated, it is hoped that the selection covered in this chapter is representative of the different types of problem that exist and includes an account of the most significant ones.

There are nine sections in this chapter, each covering a different problem. They include problems involving transversals, Cevians, the pedal triangles of a point, the pivot theorem, and the symmedians. An account is given of the Euler line and another line in the triangle in which the ratio $2:1$ also appears. Finally, we consider whether certain distances between key points in the triangle can be made integers by choosing the sides of the triangle suitably.

Areal co-ordinates are used in most of the problems, and an account and derivation of their main properties is given in the appendix. A reader not acquainted with these co-ordinates is advised to study the appendix before reading this chapter. My experience is that areal co-ordinates form a useful tool in dealing with all sorts of geometrical problems, and, as no comprehensive account of them, including how to calculate distances, appears in any modern text, I hope that the appendix proves useful.

8.1 Transversals of integer-sided triangles

In this section we consider the problem of an integer-sided triangle with given side lengths a, b, and c and ask whether it is possible to choose a transversal DEF, with D on BC, E on CA, and F on AB with the following property.

> All of the segments DE, EF, FD, BD, CD, CE, AE, AF, and BF are of rational length, so that, by enlargement, all line segments in the figure are of integer length.

It turns out that, for any given triangle with integer sides, there are an infinite number of transversals with this property.

We use areal co-ordinates, as in Section A.7 of the appendix. Suppose that the transversal has equation $lx + my + nz = 0$, where l, m, and n are integers. In order that DEF should not be parallel to any of the sides, no two of l, m, and n may be equal. Also, without loss of generality, we may assume that $m > n > 0 > l$, so that E and F are internal points of CA and AB, respectively, and D lies on BC extended beyond C. The transversal meets BC, with equation $x = 0$, at $D(0, -n, m)/(m - n)$. Similarly, E has co-ordinates $(n, 0, -l)/(n - l)$ and F has co-ordinates $(m, -l, 0)/(m - l)$.

It follows, from the discussion in Section A.2, that

$$\begin{aligned} BD &= \frac{ma}{m-n}, \quad & DC &= \frac{na}{m-n}, \quad & CE &= \frac{nb}{n-l}, \\ EA &= -\frac{lb}{n-l}, \quad & AF &= -\frac{lc}{m-l}, \quad & FB &= \frac{mc}{m-l} \end{aligned} \tag{8.1.1}$$

are all (positive) rational numbers.

We now compute the lengths of the segments FD, DE, and EF by using the areal metric. For those not familiar with the areal metric, it is given by

$$(\mathrm{d}s)^2 = -a^2\,\mathrm{d}y\,\mathrm{d}z - b^2\,\mathrm{d}z\,\mathrm{d}x - c^2\,\mathrm{d}x\,\mathrm{d}y\,, \tag{8.1.2}$$

where $\mathbf{d}\boldsymbol{r} = (\mathrm{d}x, \mathrm{d}y, \mathrm{d}z)$ is a vector displacement in the plane of the triangle, and so satisfies $\mathrm{d}x + \mathrm{d}y + \mathrm{d}z = 0$. This formula is not only true for an infinitesimal arc length, but also when $\mathbf{d}\boldsymbol{r}$ is a finite straight-line displacement. Note that the formula is true only for displacements, which is why the sum of the components of $\mathbf{d}\boldsymbol{r}$ is zero. See Section A.8 of the appendix for a derivation of the metric.

Now a short piece of algebra shows that

$$\boldsymbol{FD} = \frac{m}{(m-l)(m-n)}(-m+n, -n+l, -l+m)\,.$$

It follows from eqn (8.1.2) that

$$\begin{aligned} (m-l)^2(m-n)^2\frac{FD^2}{m^2} &= a^2l^2 + b^2m^2 + c^2n^2 - lm(a^2+b^2-c^2) \\ &\quad - mn(b^2+c^2-a^2) - nl(c^2+a^2-b^2)\,. \end{aligned} \tag{8.1.3}$$

By symmetry, this is also equal to

$$(n-m)^2(n-l)^2\frac{DE^2}{n^2} = (l-n)^2(l-m)^2\frac{EF^2}{l^2}\,.$$

If, for the moment, we put the term on the right-hand side of eqn (8.1.3) equal to d^2, where d is positive, then, in our figure, in which $FD = EF + DE$, we have

$$\begin{aligned} FD &= \frac{md}{(m-l)(m-n)}, \\ EF &= -\frac{ld}{(n-l)(m-l)}, \\ DE &= \frac{nd}{(m-n)(n-l)}\,. \end{aligned} \tag{8.1.4}$$

Essentially, this means that, if any one of the segments FD, EF, and DE is rational, then so are the other two.

Suppose now that we are given an integer-sided triangle ABC. Then, provided that we can choose l, m, and n so that d is an integer for any integers a, b, and c, it follows that all nine line segments in the configuration of the triangle will be rational.

Theorem 8.1.1 *If a, b, and c are the integer sides of a triangle, then there exist an infinite number of sets of integers l, m, n, and d (depending on a, b, and c) such that*

$$a^2l^2 + b^2m^2 + c^2n^2 - mn(b^2 + c^2 - a^2) - nl(c^2 + a^2 - b^2) - lm(a^2 + b^2 - c^2) = d^2 . \tag{8.1.5}$$

Hence, by virtue of eqns (8.1.1) and (8.1.4), there are an infinite number of transversals DEF of the triangle ABC, with equations $lx + my + nz = 0$ in areal co-ordinates, such that the line segments BD, CD, CE, AE, AF, BF, DE, EF, and FD are all of rational length.

Proof The method of solving eqn (8.1.5) is the same as in Sections 1.2 and 7.5; that is, we find a non-singular linear change of variable between l, m, n, and d, and u, v, w, and e such that eqn (8.1.5) becomes linear in each variable. A suitable change of variable is

$$l = u + e\,, \qquad m = v + e\,, \qquad n = w + e\,, \qquad d = au + bv + cw\,. \tag{8.1.6}$$

Substituting into eqn (8.1.5), it turns out that this equation is satisfied provided that we choose

$$w = -\frac{(a + b - c)uv}{(c + a - b)u + (b + c - a)v}\,. \tag{8.1.7}$$

Substituting back into eqn (8.1.6) and multiplying up by the denominator of w, we find the following three-parameter solution for l, m, n, and d in terms of e, u, and v:

$$\begin{aligned} l &= (u + e)[(c + a - b)u + (b + c - a)v]\,, \\ m &= (v + e)[(c + a - b)u + (b + c - a)v]\,, \\ n &= -(a + b - c)uv + e[(c + a - b)u + (b + c - a)v]\,, \\ d &= a(c + a - b)u^2 + (c + a - b)(c + b - a)uv + b(c + b - a)v^2\,. \end{aligned} \tag{8.1.8}$$

Since e, u, and v may have any integer values such that $m > n > 0 > l$, the theorem is proved. □

For example, with $a = 3$, $b = 4$, $c = 5$, $u = -3$, $e = 2$, and $v = 4$, then after cancelling common factors we obtain $l = -1$, $m = 6$, $n = 4$, and $d = 17$. We now obtain $BD = 9$, $DC = 6$, $CE = 16/5$, $EA = 4/5$, $AF = 5/7$, $FB = 30/7$, $FD = 51/7$, $FE = 17/35$, and $ED = 34/5$. Enlargement by a factor of 35 makes all nine line segments integer in length, see Fig. 8.1.

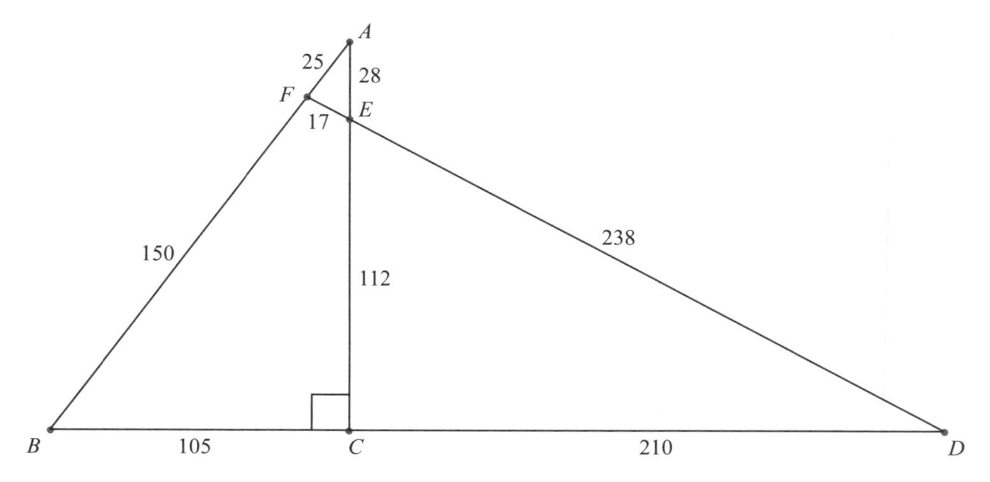

Fig. 8.1 An integer-sided triangle with integer segments on a transversal.

Exercises 8.1

8.1.1 Use the parameters $u = -3$, $e = 1$, and $v = 3$ in eqns (8.1.8) for the triangle with $a = 2$, $b = 3$, and $c = 4$ to find a transversal such that all of the line segments in the configuration are rational.

8.1.2 Repeat Exercise 8.1.1 with $u = -1$, $e = 0$, and $v = 1$ for the triangle with $a = 2$, $b = 3$, and $c = 4$.

8.1.3 Simplify eqns (8.1.8) for an equilateral triangle of side a and construct an example of the theory.

8.2 The pedal triangle of three Cevians

Suppose that ABC is an integer-sided triangle and P is a point in the plane of the triangle, not lying on any of the sides. Let AP meet BC at L, BP meet CA at M, and CP meet AB at N. The three lines AL, BM, and CN have become known as *Cevians*.

This is because of Ceva's theorem, which states that under these circumstances

$$\frac{BL}{LC}\frac{CM}{MA}\frac{AN}{NB} = 1\,. \tag{8.2.1}$$

In eqn (8.2.1) the lengths are signed, so that, for example, if L is on the extension of BC beyond C then BL is positive but LC is negative. Ceva's theorem is proved in Section A.7 of the appendix. The triangle LMN is called *the pedal triangle of the three Cevians*. There are three problems. Firstly, we consider whether it is possible to find points P such that, in suitably chosen integer-sided triangles, the sides of the pedal triangle are also integers. There is also the question of whether there are points

P for which LMN is integer-sided for all or most integer-sided triangles. The third problem is to find the points P for which LMN is integer-sided when the sides of the triangle are specified integers.

Most of these problems are very difficult, but answers are known for certain points P. One case can be solved straight away. When P is at G, the centroid, the triangle LMN is the median triangle, whose sides are parallel to those of ABC and half their length. So LMN is integer-sided if and only if ABC has even integer sides.

For other cases it is necessary to work out LM, MN, and NL for a triangle of integer side lengths a, b, and c for a point P with areal co-ordinates (l, m, n). We restrict our discussion to cases in which P is internal to the triangle ABC.

Let the areal co-ordinates of P be (l, m, n), where l, m, and n are rational numbers lying strictly between 0 and 1, such that $l + m + n = 1$. Then L, M, and N have co-ordinates

$$L\frac{(0,m,n)}{m+n}, \quad M\frac{(l,0,n)}{n+l}, \quad N\frac{(l,m,0)}{l+m}.$$

It follows that

$$\boldsymbol{LM} = \frac{(l(m+n), -m(n+l), n(m-l))}{(n+l)(m+n)}.$$

Using the areal metric, given by eqn (8.1.2) and established in Section A.8 of the appendix, we obtain

$$(n+l)^2(n+m)^2LM^2 = a^2mn(n+l)(m-l) - b^2nl(m+n)(m-l) + c^2lm(m+n)(n+l).$$

Putting $l = 1/u$, $m = 1/v$, and $n = 1/w$, where u, v, and w are rational numbers strictly greater than 1 and which, since $l + m + n = 1$, satisfy

$$uv + vw + wu = uvw, \tag{8.2.2}$$

we obtain

$$(w+u)^2(w+v)^2\frac{LM^2}{w^2} = a^2u^2 + b^2v^2 + c^2w^2 + vw(b^2 + c^2 - a^2) + wu(c^2 + a^2 - b^2) - uv(a^2 + b^2 - c^2). \tag{8.2.3}$$

Similarly,

$$(u+v)^2(u+w)^2\frac{MN^2}{u^2} = a^2u^2 + b^2v^2 + c^2w^2 - vw(b^2 + c^2 - a^2) + wu(c^2 + a^2 - b^2) + uv(a^2 + b^2 - c^2) \tag{8.2.4}$$

and

$$(v+w)^2(v+u)^2\frac{NL^2}{v^2} = a^2u^2 + b^2v^2 + c^2w^2 + vw(b^2 + c^2 - a^2) - wu(c^2 + a^2 - b^2) + uv(a^2 + b^2 - c^2). \tag{8.2.5}$$

Since the three expressions for LM^2, MN^2, and NL^2 are homogeneous of degree 0 in u, v, and w, we may take u, v, and w to be integers and discard the normalisation equation (8.2.2). For, given any rational solution of eqn (8.2.2), we can multiply each of u, v, and w by the least common multiple of their denominators to obtain integer u, v, and w, and the values of LM, MN, and NL are unaltered. Conversely, any set of integers u, v, and w can be scaled to provide a rational solution of eqn (8.2.2).

If we refer back to Section 8.1, then we see that the expression for NL^2 is the same as that for FD^2 under the correspondence $v \to -m$, $u \to l$, and $w \to n$. This means that it is very easy to make one or other of LM, MN, and NL rational by using a correspondingly modified parameter system. The problem of making all of LM, MN, and NL rational for any given triangle requires making three homogeneous quadratic forms simultaneously perfect squares, a problem analogous to that of finding integer-sided triangles with three integer medians. We now consider a number of special cases.

- *Case 1*
  For any triangle there is the trivial solution with $u = v = w = 3$, which gives the median triangle in which $MN = \frac{1}{2}a$, $NL = \frac{1}{2}b$, and $LM = \frac{1}{2}c$. So, whenever a, b, and c are even, LMN has integer sides. This corresponds to $l = m = n = 1/3$ and P is the centroid G. This case has already been mentioned above.
- *Case 2*
  Suppose that the triangle ABC is equilateral with side a. Then eqns (8.2.3)–(8.2.5) become

$$(w+u)^2(w+v)^2LM^2 = a^2w^2(u^2+v^2+w^2+vw+wu-uv)\,, \qquad (8.2.6)$$
$$(u+v)^2(u+w)^2MN^2 = a^2u^2(u^2+v^2+w^2-vw+wu+uv)\,, \qquad (8.2.7)$$
$$(v+w)^2(v+u)^2NL^2 = a^2v^2(u^2+v^2+w^2+vw-wu+uv)\,. \qquad (8.2.8)$$

  This illustrates what was stated above, that searching for points P for any particular triangle, in general, requires making three homogeneous quadratic expressions simultaneously perfect squares. It is certain that an infinite number of sets of values of u, v, and w exist such that each of LM, MN, and NL is rational. For example, there are an infinite number of solutions for an isosceles triangle when P lies on the line of symmetry, see Case 6 below.
- *Case 3*
  For an acute integer-sided triangle, it is always possible to find an internal point other than the centroid such that the pedal triangle of the Cevians has rational sides. This is when the internal point P is the orthocentre H. Then (l, m, n) is proportional to $(\tan A, \tan B, \tan C)$. It is known, in this case, that

$$MN = a\cos A\,, \quad NL = b\cos B\,, \quad LN = c\cos C\,. \qquad (8.2.9)$$

  These lengths are rational for any acute integer-sided triangle. Note that l, m, and n are rational only when the triangle is a Heron triangle, showing that a case certainly exists in which the sides of the pedal triangle are rational and l, m, and n are not.

In eqns (8.2.3)–(8.2.5) the following terms are equal:

$$vw(b^2 + c^2 - a^2) = wu(c^2 + a^2 - b^2) = uv(a^2 + b^2 - c^2), \qquad (8.2.10)$$

since each is proportional to $\cos A \cos B \cos C$. So, as soon as one of the expressions (8.2.3), (8.2.4), or (8.2.5) is a perfect square, then so are the other two expressions. As an example, take $a = 14$, $b = 13$, $c = 15$, $\tan A = 56/33$, $\tan B = 4/3$, and $\tan C = 12/5$. The normalised areal co-ordinates of H are $(\cot B \cot C, \cot C \cot A, \cot A \cot B)$. It follows that $u = 16/5$, $v = 224/55$, and $w = 224/99$. The expression on the right-hand side of each of eqns (8.2.3)–(8.2.5) is then $(2912/33)^2$. It follows that $MN = 462/65$, $NL = 39/5$, and $LM = 75/13$. The total perimeter is $1344/65$, see Fig. 8.2. The pedal triangle of the orthocentre of an acute-angled triangle is of interest geometrically because its boundary is the shortest path that a light ray can traverse a closed path reflecting off the sides of the triangle as if they are mirrors. This is because the normals at L, M, and N bisect the angles of the triangle LMN, which are $180° - 2A$, $180° - 2B$, and $180° - 2C$.

- *Case 4*

  Another example of a pedal triangle with rational sides is when the sines and cosines of all of the angles $\frac{1}{2}A$, $\frac{1}{2}B$, and $\frac{1}{2}C$ are rational, and the internal point P is Gergonne's point. If X, Y, and Z are the points where the incircle touches the sides BC, CA, and AB, respectively, then AX, BY, and CZ are concurrent at Gergonne's point, with unnormalised areal co-ordinates $(1/(s-a), 1/(s-b), 1/(s-c))$, see Section A.7 of the appendix for a derivation. The values of u, v, and w may therefore be taken as $s - a$, $s - b$, and $s - c$, respectively. However, it is known from the geometrical configuration, without any need to use eqns (8.2.3)–(8.2.5), that $YZ = (b+c-a)\sin\frac{1}{2}A$, $ZX = (c+a-b)\sin\frac{1}{2}B$, and $XY = (a+b-c)\sin\frac{1}{2}C$, see Exercise 8.2.1.

- *Case 5*

  An open question is whether, given a point with fixed rational areal co-ordinates, there is always a triangle with a pedal triangle, with respect to that point, having

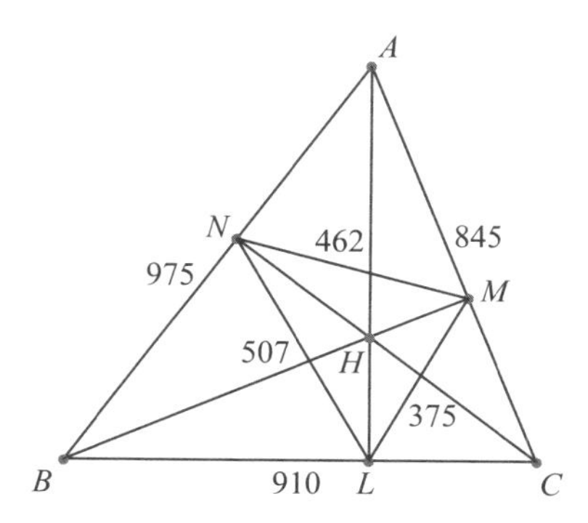

Fig. 8.2 An integer-sided triangle with a pedal triangle having integer sides.

rational side lengths. For example, when $u = 2$, $v = 3$, and $w = 1$, this amounts to finding integer a, b, and c such that all of $-3a^2 + 4b^2 + 12c^2$, $15a^2 + 10b^2 - 6c^2$, and $5a^2 + 20b^2 - 4c^2$ are perfect squares. For each set of values of u, v, and w this presents a problem similar to that of finding integer-sided triangles in which all of the medians are of integer length. Experience with that problem leads me to suspect that such triangles may exist for some choices of u, v, and w, but, unless it can be proved (or disproved) that such is the case for *all* choices of u, v, and w, the matter is not very interesting.

- *Case 6*

  When the triangle is integer-sided and isosceles it turns out that there are an infinite number of points on the line of symmetry, with rational areal co-ordinates, leading to pedal triangles with rational sides.

  To see this, let us put $c = b$ and $w = v$ in expressions (8.2.3)–(8.2.5). We find that $LM = NL$, and from eqn (8.2.3) their lengths are given by

$$4(u+v)^2 LM^2 = a^2u^2 + (4b^2 - a^2)v^2 \,. \tag{8.2.11}$$

From eqn (8.2.4) we have

$$(u+v)^4 \frac{MN^2}{u^2} = a^2(u+v)^2 \,, \tag{8.2.12}$$

that is, $MN = au/(u+v)$. In other words, provided that u and v are rational, MN is automatically rational. This, of course, is because the triangles ANM and ABC are similar. We now show that, whatever values a and b have, it is always possible to find infinitely many pairs of integers u and v such that

$$a^2u^2 + (4b^2 - a^2)v^2 = x^2 \,, \tag{8.2.13}$$

where x is integral. We make a non-singular linear change of variable so that, in terms of the new variables, eqn (8.2.13) becomes linear in each variable. A suitable change of variable is

$$u = p + q + r \,, \quad v = p - r \,, \quad x = 2b(p+r) + aq \,. \tag{8.2.14}$$

Then eqn (8.2.13) gives $q = -2(a+2b)pr/a(p+r)$. Substituting this value back into eqns (8.2.14) and multiplying up by $a(p+r)$, we obtain

$$u = a(p^2+r^2) - 4bpr \,, \quad v = a(p^2-r^2) \,, \quad x = 2ab(p^2+r^2) - 2a^2pr \,. \tag{8.2.15}$$

In eqn (8.2.15) p and r may have any integer values such that u, v, and w are positive. Any common factors may be extracted. As an example, let $a = 7$, $b = 5$, $p = 3$, and $r = 1$. Then $u = 5$, $v = 28$, and $x = 203$. It follows that $MN = 35/33$ and $NL = LM = 203/66$.

Exercises 8.2

8.2.1 Find the sides of the pedal triangle formed by the feet of the Cevians through Gergonne's point in the triangle with $a = 154$, $b = 169$, and $c = 125$.

8.2.2 Show that, if u, v, and w are fixed positive integers and $v = w$, then there exist infinitely many isosceles triangles with $b = c$, such that the pedal triangle LMN has rational side lengths.

8.3 The pedal triangle of a point

If P is a point in the plane of a triangle ABC not lying on any of its sides and the feet of the perpendiculars from P onto the sides BC, CA, and AB are denoted by L, M, and N, respectively, then the triangle LMN is called *the pedal triangle of the point* P.

Similar problems exist as those detailed in Section 8.2 for the pedal triangle of the Cevians. We may ask whether it is possible to find points P such that, in suitably chosen integer-sided triangles, the sides of the pedal triangle of P are also integers. There is also the question of whether there are points P for which LMN is an integer-sided triangle for all or most integer-sided triangles. The third problem is to find points P for which LMN is an integer-sided triangle when the sides of the triangle ABC are specified.

In what follows we restrict P to be an internal point and suppose that it has areal co-ordinates (l, m, n), where l, m, and n are rational numbers lying strictly between 0 and 1 and are such that $l + m + n = 1$.

Three lines through P are drawn perpendicular to the sides BC, CA, and AB and meet them at L, M, and N, respectively. Then, as shown in Section A.9 of the appendix,

$$BL = an + cl\cos B\,, \tag{8.3.1}$$

$$NB = cl + an\cos B\,. \tag{8.3.2}$$

So, by the cosine rule for the triangle NBL, we have

$$NL^2 = BL^2 + NB^2 - 2\,BL\,NB\cos B = \sin^2 B(a^2n^2 + c^2l^2 + 2canl\cos B)\,. \tag{8.3.3}$$

Similarly,

$$LM^2 = \sin^2 C(b^2l^2 + a^2m^2 + 2ablm\cos C) \tag{8.3.4}$$

and

$$MN^2 = \sin^2 A(c^2m^2 + b^2n^2 + 2bcmn\cos A)\,. \tag{8.3.5}$$

- *Case 1*

  Let P be the circumcentre O; then L, M, and N are the midpoints of the sides. The above formulae are not needed. LMN is the median triangle and $MN = \frac{1}{2}a$,

$NL = \frac{1}{2}b$, and $LM = \frac{1}{2}c$, and so LMN is integer-sided if and only if ABC has sides that are even integers.

- *Case 2*
Suppose now that the triangle ABC is equilateral with side a. Then $\sin^2 A = \sin^2 B = \sin^2 C = 3/4$ and $\cos A = \cos B = \cos C = 1/2$. It follows from eqns (8.3.3)–(8.3.5) that we have

$$\begin{aligned} NL^2 &= \frac{3}{4}a^2(n^2 + nl + l^2)\,, \\ LM^2 &= \frac{3}{4}a^2(l^2 + lm + m^2)\,, \\ MN^2 &= \frac{3}{4}a^2(m^2 + mn + n^2)\,. \end{aligned} \tag{8.3.6}$$

Hence NL, LM, and MN are rational whenever rational numbers l, m, and n are found so that $l + m + n = 1$ and each of $n^2 + nl + l^2$, $l^2 + lm + m^2$, and $m^2 + mn + n^2$ are 3 times a rational square. Then a can be chosen so that NL, LM, and MN are integers. This problem is very similar to the problem of the equilateral triangle treated in detail in Section 6.4. A computer search has been carried out by J. T. Bradley (private communication) for finding integer values of l, m, and n such that all of n^2+nl+l^2, l^2+lm+m^2, and m^2+mn+n^2 are 3 times a perfect square, and for which $10\,000 > l > m > n > 0$. The strict inequalities here mean that the corresponding solutions do not lie on any of the medians, as solutions for points on the medians may be obtained for the case of the isosceles triangle, see Case 6 below. The results of the computer search are given in Table 8.1. Solutions are few and far between as compared with medians of integer length for integer-sided triangles.

- *Case 3*
For any acute integer-sided triangle, NL, LM, and MN are rational when P is the orthocentre H. Then $l = \cot B \cot C$, $m = \cot C \cot A$, and $n = \cot A \cot B$ (see Section A.6 of the appendix). Putting $a = 2R \sin A$, etc., we have

Table 8.1 Integer values of l, m, and n such that each of $n^2 + nl + l^2$, $l^2 + lm + m^2$, and $m^2 + mn + n^2$ are 3 times a perfect square.

l	m	n
611	338	83
1261	598	253
2522	1067	506
2714	1898	803
3674	2714	803
6578	2783	386
6887	5522	3266
8327	5447	1514

$$NL^2 = 4R^2\cos^2 B(\cos^2 A + \cos^2 C + 2\cos A\cos B\cos C)$$
$$= 4R^2\cos^2 B(1 - \cos^2 B),$$

and hence $NL = 2R\cos B\sin B = b\cos B$. Similarly, $LM = c\cos C$ and $MN = a\cos A$. This is, of course, the same result as in Section 8.2, since for the orthocentre the perpendiculars through H coincide with the altitudes AD, BE, and CF, which are the Cevians of the point H.

- *Case 4*
  When P is the incentre I, the feet of the perpendicular are the points X, Y, and Z where the incentre touches the sides BC, CA, and AB, respectively. So the pedal triangle XYZ coincides with the pedal triangle of the Cevians of Gergonne's point. Hence there are rational sides when the sines and cosines of $\frac{1}{2}A$, $\frac{1}{2}B$, and $\frac{1}{2}C$ are all rational.
- *Case 5*
  We now consider the case of an isosceles triangle. When $c = b$ and $n = m$ the formulae (8.3.3)–(8.3.5) become

$$NL^2 = LM^2 = \sin^2 B(a^2m^2 + b^2l^2 + 2ablm\cos B) \tag{8.3.7}$$

and

$$MN^2 = 2b^2m^2\sin^2 A(1 + \cos A) = 4b^2m^2\sin^2 A\cos^2\frac{1}{2}A. \tag{8.3.8}$$

Hence $MN = 2bm\sin A\cos\frac{1}{2}A$, which is rational provided that $\cos\frac{1}{2}A$ is rational. (Since $a = 2b\sin\frac{1}{2}A$, we already know that $\sin\frac{1}{2}A$ is rational.) All of the sides of the pedal triangle will then be rational provided that we can choose l and m so that

$$a^2m^2 + b^2l^2 + 2ablm\cos B \tag{8.3.9}$$

is the square of a rational number and $\sin B$ is rational. Now $\cos B = a/2b$, so expression (8.3.9) becomes $a^2m^2 + b^2l^2 + a^2lm$.

A two-parameter solution of the Diophantine equation $a^2m^2 + b^2l^2 + a^2lm = x^2$ is

$$\begin{aligned} l &= 4apr\,, \\ m &= (r-p)[(2b-a)p + (2b+a)r]\,, \\ x &= a(2b-a)p^2 + a(2b+a)r^2\,. \end{aligned} \tag{8.3.10}$$

In eqn (8.3.10) p and r are any integers such that l and m are positive. It is worth seeing how to use the solution, as integral values for l and m, once obtained, have then to be normalised so that $l + 2m = 1$, and it is also necessary that $\sin B$ should be made rational.

As an example, let us take the isosceles triangle with $a = 10$ and $b = c = 13$. Then $\sin\frac{1}{2}A = \cos B = 5/13$ and $\cos\frac{1}{2}A = \sin B = 12/13$. Let us choose $r = 5$ and $p = 2$ in eqns (8.3.10). Then $l = 400$, $m = 636$, and $x = 9640$. Now the expressions that we are dealing with are all homogeneous quadratics in l and m, so we may normalise by dividing all of l, m, and x by $l + 2m = 1672$. This means that their actual values are $l = 50/209$, $m = n = 159/418$, and $x = 1205/209$. We can now finish the calculation to obtain $NL = LM = x \sin B = 14\,460/2717$ and $MN = 2bm \sin A \cos \frac{1}{2}A = 228\,960/35\,321$.

Exercise 8.3

8.3.1 Take the isosceles triangle with $a = 6$ and $b = c = 5$, and the parameters $r = 3$ and $p = 1$ in eqns (8.3.10), and find the side lengths of the corresponding pedal triangle.

8.4 The pivot theorem

Firstly, we define the pivot point P. We then show that the areal co-ordinates of P are rational provided that the points L, M, and N that create the pivot point also have rational areal co-ordinates.

Theorem 8.4.1 *If ABC is a triangle and the points L, M, and N are chosen on the lines BC, CA, and AB, respectively, but not at the vertices, then the circles AMN, BNL, and CLM meet at a point P, called the* pivot point.

Proof We give the proof only for the case when P lies inside the triangle ABC, but the proof is much the same when P lies outside the triangle ABC. Let the circles BNL and CLM meet at P. Since $BNPL$ is cyclic we have $\angle NPL = 180° - B$, and since $CMPL$ is cyclic we have $\angle MPL = 180° - C$. It follows that $\angle MPN = 360° - (180° - B) - (180° - C) = 180° - A$. Hence $AMPN$ is cyclic; that is, the circle AMN passes through P. □

As stated above, the problem which we solve in this section is to find the ratios $[BCP] : [CAP] : [ABP] : [ABC] = x : y : z : x + y + z$, given the points L, M, and N. Suppose that $BL/LC = (1 - l)/l$, $CM/MA = (1 - m)/m$, and $AN/NB = (1 - n)/n$. We determine the values of x, y, and z, which are the unnormalised areal co-ordinates of P, in terms of the six parameters a, b, c, l, m, and n, showing them to be rational, provided that the parameters l, m, and n are rational and ABC is an integer-sided triangle.

The areal co-ordinates of L, M, and N are $(0, l, 1 - l)$, $(1 - m, 0, m)$, and $(n, 1 - n, 0)$, respectively. We also use the fact that the conic with equation

$$ux^2 + vy^2 + wz^2 + 2pyz + 2qzx + 2rxy = 0 \tag{8.4.1}$$

is a circle if and only if

$$\frac{v + w - 2p}{a^2} = \frac{w + u - 2q}{b^2} = \frac{u + v - 2r}{c^2}. \tag{8.4.2}$$

See Section A.10 of the appendix for the proof. In view of the homogeneity of eqn (8.4.1), we may as well take the common value in eqn (8.4.2) to be 1. We now put in the conditions that the circle (8.4.1) passes through A, M, and N. These conditions are

$$\begin{aligned} u &= 0\,, \\ wm + 2qm(1-m) &= 0\,, \\ v(1-n) + 2rn &= 0\,. \end{aligned} \tag{8.4.3}$$

Solving these equations we find the circle AMN to have equation

$$c^2ny^2+b^2(1-m)z^2+[c^2n+b^2(1-m)-a^2]yz-b^2mzx-c^2(1-n)xy=0\,. \tag{8.4.4}$$

Similarly, the circles BNL and CLM have equations

$$a^2lz^2+c^2(1-n)x^2+[a^2l+c^2(1-n)-b^2]zx-c^2nxy-a^2(1-l)yz=0 \tag{8.4.5}$$

and

$$b^2mx^2+a^2(1-l)y^2+[b^2m+a^2(1-l)-c^2]xy-a^2lyz-b^2(1-m)zx=0\,, \tag{8.4.6}$$

respectively.

If x, y, and z are normalised areal co-ordinates of P then $x+y+z=1$ and eqns (8.4.5) and (8.4.6) simplify to

$$a^2lz - c^2nx + c^2x^2 + c^2zx - b^2zx - a^2yz = 0 \tag{8.4.7}$$

and

$$b^2mx - a^2ly + a^2y^2 + a^2xy - c^2xy - b^2zx = 0\,, \tag{8.4.8}$$

respectively. We next subtract these equations and use $x+y+z=1$ again to obtain

$$[b^2m - c^2(1-n)]x + a^2(1-l)y - a^2lz = 0\,. \tag{8.4.9}$$

A cyclic change of a, b, and c; l, m, and n; and x, y, and z gives a second linear equation

$$-b^2mx + [c^2n - a^2(1-l)]y + b^2(1-m)z = 0\,. \tag{8.4.10}$$

Since we are only interested in ratios, we can now relax the condition $x+y+z=1$ and use these two equations to derive (unnormalised) values for x, y, and z. These are

$$\begin{aligned} x &= a^2\left[(1-l)(1-m)b^2 + nlc^2 - l(1-l)a^2\right], \\ y &= b^2\left[(1-m)(1-n)c^2 + lma^2 - m(1-m)b^2\right], \\ z &= c^2\left[(1-n)(1-l)a^2 + mnb^2 - n(1-n)c^2\right]. \end{aligned} \tag{8.4.11}$$

Exercises 8.4

8.4.1 Where is the pivot point when L, M, and N are (a) the feet of the medians, and (b) the feet of the altitudes?

8.4.2 Find $[BCP] : [CAP] : [ABP]$, where P is the pivot point and L, M, and N are the feet of the internal bisectors.

8.4.3 Invert the pivot theorem about any point to establish the six-circle theorem. This states that, if C_1, C_2, C_3, and C_4 are four circles, and if the intersections of C_j and C_k are denoted by P_{jk} and Q_{jk} ($j, k = 1, 2; 2, 3; 3, 4; 4, 1$), then the four points Q_{jk} are concyclic if and only if the four points P_{jk} are concyclic.

8.5 The symmedians and other Cevians

We now prove a general theorem linking the length of the Cevians of a point X with the length of the Cevians of a point Y, where Y is the isogonal conjugate of X. For the definition of the isogonal conjugate of a point see Section A.10 of the appendix. The theorem tells us that, if the triangle ABC is integer-sided and if one set of Cevians are rational in length, then the other set of Cevians are rational in length. In fact, the lengths are related so that, if the Cevian AXL is rational, then so is the Cevian AYP, where L and P lie on BC. We begin by looking at the particular case when X is the centroid and Y is the symmedian or Lemoine point.

The *symmedian point* S in the triangle ABC has areal co-ordinates $(a^2, b^2, c^2)/(a^2 + b^2 + c^2)$. It is the isogonal conjugate of the centroid G, and has the following interesting properties.

(i) If AL, BM, and CN are the medians, AU, BV, and CW are the internal bisectors, and AP, BQ, and CR are the symmedians, then the angles LAP, MBQ, and NCR are bisected by the internal bisectors AU, BV, and CW, respectively.

(ii) If the tangents to the circumcircle ABC at B and C meet at D, and E and F are similarly defined, then AD, BE, and CF are concurrent at S.

Proof of part (i) is not given here, but appears in Bradley (2005). Part (ii) is proved in Section A.10 of the appendix.

Using the areal metric, defined by eqn (8.1.2), we have

$$AP^2 = \frac{b^2c^2(2b^2 + 2c^2 - a^2)}{(b^2 + c^2)^2}. \tag{8.5.1}$$

It follows that

$$AP = \frac{2bcAL}{b^2 + c^2}. \tag{8.5.2}$$

Hence, in a triangle that is not isosceles, a symmedian is shorter than the corresponding median. Also, for an integer-sided triangle, the length of a symmedian is rational if and only if the length of the corresponding median is rational. A list of many integer-sided triangles in which all three medians are integers is given in Section 2.4, so the results of that section apply here also, to the case when all three symmedians are of rational length. The above property is a particular case of a more general theorem.

Theorem 8.5.1 *Suppose that X is a point in the plane of an integer-sided triangle ABC with rational areal co-ordinates (l, m, n). Let AL, BM, and CN be the Cevians through X. Suppose that Y is the isogonal conjugate of X, with areal co-ordinates $(a^2/l, b^2/m, c^2/n)$. Also, let AP, BQ, and CR be the Cevians through Y. Then AP is rational in length if and only if AL is rational in length, and similarly for the pairs BQ and BM, and CR and CN.*

Proof Using the areal metric, given by eqn (8.1.2), we have

$$AL^2 = \frac{(b^2 + c^2 - a^2)mn + b^2n^2 + c^2m^2}{(m+n)^2}. \tag{8.5.3}$$

Also,

$$AP^2 = \frac{b^2c^2[(b^2 + c^2 - a^2)mn + b^2n^2 + c^2m^2]}{(b^2n + c^2m)^2}, \tag{8.5.4}$$

and hence

$$AP = \frac{bc(m+n)AL}{b^2n + c^2m}. \tag{8.5.5}$$

The rest of the proof follows by cyclic changes of letters. □

Exercises 8.5

8.5.1 Find the lengths of the symmedians in the triangle with $a = 136$, $b = 170$, and $c = 174$.

8.5.2 In the triangle with $a = 14$, $b = 13$, and $c = 15$, calculate the length of the line AOP, where O is the circumcentre and P lies on BC. (Use the fact that O is the isogonal conjugate of H, the orthocentre.)

8.6 The Euler line and ratios $2 : 1$ in a triangle

In this section we give a proof of the fact that $OGTH$ is a straight line, where O is the circumcentre, G is the centroid, T is the centre of the nine-point circle, and H is the orthocentre of the triangle ABC, and establish the relations $\boldsymbol{OG} = (1/3)\boldsymbol{OH}$ and $\boldsymbol{OT} = (1/2)\boldsymbol{OH}$. The line $OGTH$ is called the *Euler line*. We also identify another line in the triangle with four points on it, which has similar properties as regards the ratios of the distances between the points.

The Euler line

In the triangle ABC denote the feet of the altitudes by D, E, and F and the midpoints of the sides by L, M, and N. If O is the circumcentre, then $\boldsymbol{OL} = \frac{1}{2}(\boldsymbol{OB}+\boldsymbol{OC})$. Define the point H by the equation $\boldsymbol{AH} = 2\boldsymbol{OL}$. Since OL is perpendicular to BC, AH is also perpendicular to BC, and hence H lies on the altitude AD. The vector position of H is given by $\boldsymbol{OH} = \boldsymbol{OA}+\boldsymbol{AH} = \boldsymbol{OA}+\boldsymbol{OB}+\boldsymbol{OC}$. By symmetry, H also lies on the altitudes BE and CF. H is therefore the point where the three altitudes meet. It is therefore the orthocentre of ABC. The point G on AL such that $AG : GL = 2 : 1$ has the vector position $\boldsymbol{OG} = \boldsymbol{OA} + (2/3)(\boldsymbol{OL} - \boldsymbol{OA}) = (1/3)(\boldsymbol{OA} + \boldsymbol{OB} + \boldsymbol{OC})$. Hence G lies on OH and $GH : OG = 2 : 1$. Since $\angle AGH = \angle LGO$, it follows that the triangles AGH and LGO are similar, with a scale factor of 2.

The nine-point circle

Let T be the midpoint of OH, so that $\boldsymbol{OT} = \frac{1}{2}(\boldsymbol{OA} + \boldsymbol{OB} + \boldsymbol{OC})$. Denote the midpoints of AH, BH, and CH by U, V, and W, respectively. We have $\boldsymbol{OU} = \boldsymbol{OA}+\frac{1}{2}(\boldsymbol{OB}+\boldsymbol{OC})$, $\boldsymbol{OV} = \boldsymbol{OB}+\frac{1}{2}(\boldsymbol{OC}+\boldsymbol{OA})$, and $\boldsymbol{OW} = \boldsymbol{OC}+\frac{1}{2}(\boldsymbol{OA}+\boldsymbol{OB})$. Also, $\boldsymbol{OL} = \frac{1}{2}(\boldsymbol{OB} + \boldsymbol{OC})$, $\boldsymbol{OM} = \frac{1}{2}(\boldsymbol{OC} + \boldsymbol{OA})$, and $\boldsymbol{ON} = \frac{1}{2}(\boldsymbol{OA} + \boldsymbol{OB})$. Clearly, T is equidistant from U, V, W, L, M, and N, since its vector position is $\boldsymbol{OT} = \frac{1}{2}(\boldsymbol{OA} + \boldsymbol{OB} + \boldsymbol{OC})$, and so $|\boldsymbol{TU}| = |\boldsymbol{TV}| = |\boldsymbol{TW}| = |\boldsymbol{TL}| = |\boldsymbol{TM}| = |\boldsymbol{TN}| = \frac{1}{2}R$, where R is the circumradius. Now $HOLD$ is a right-angled trapezium, and, since T is the midpoint of OH, it follows that $TD = TL$. Similarly, $TE = TM$ and $TF = TN$. Hence T is the centre of a circle of radius $\frac{1}{2}R$ passing through L, M, N, U, V, W, D, E, and F. This circle is called the *nine-point circle*. Note also that $OG : GT = 2 : 1$. Since the ratio of the radius of the circumcircle to the radius of the nine-point circle is $2 : 1$, it follows that H is the external centre of similitude of the two circles and G is the internal centre of similitude of the two circles. This means, for example, that $HX = 2HD$, where X is the point at which HD meets the circumcircle. See Fig. 8.3 for a diagram of this configuration.

Another line with $2 : 1$ ratios

In this section we again use areal co-ordinates. The incentre I has co-ordinates $(a, b, c)/(a + b + c)$, see Section A.6 of the appendix, and G has co-ordinates $(1/3, 1/3, 1/3)$. The centre of mass J of a framework of uniform wires of equal density forming the sides of the triangle ABC has co-ordinates

$$\frac{a(0, \frac{1}{2}, \frac{1}{2}) + b(\frac{1}{2}, 0, \frac{1}{2}) + c(\frac{1}{2}, \frac{1}{2}, 0)}{a+b+c} = \frac{1}{2}\frac{(b+c, c+a, a+b)}{a+b+c}.$$

It follows that $\boldsymbol{IG} = (b + c - 2a, c + a - 2b, a + b - 2c)/3(a + b + c)$ and $\boldsymbol{GJ} = (b + c - 2a, c + a - 2b, a + b - 2c)/6(a + b + c)$. Hence G lies on IJ and satisfies $IG : GJ = 2 : 1$.

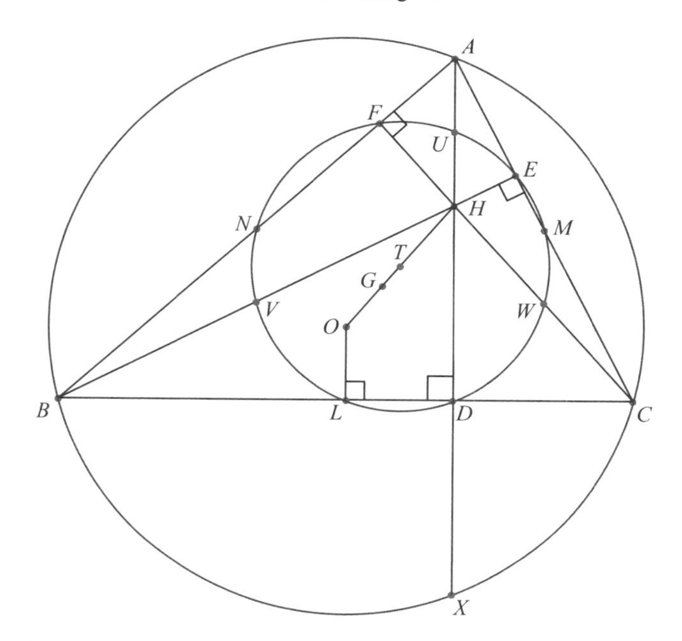

Fig. 8.3 The Euler line and the nine-point circle $DLVNFUEMW$.

Consider the escribed circle opposite A, and suppose that it touches BC at P. Then $c + BP = b + CP$ and $BP + CP = a$. It follows that $BP = \frac{1}{2}(a + b - c)$ and $CP = \frac{1}{2}(a - b + c)$. Hence the areal co-ordinates of P are $(0, a - b + c, a + b - c)/2a$. It follows that AP passes through the point W with co-ordinates $(b + c - a, c + a - b, a + b - c)/(a + b + c)$. By symmetry, BQ and CR also pass through W, where Q is the point at which the escribed circle opposite B touches CA and R is the point at which the escribed circle opposite C touches AB. W is called *Nagel's point*. Now $\boldsymbol{IW} = (b + c - 2a, c + a - 2b, a + b - 2c)/(a + b + c)$. Hence W lies on the extension of IGJ and $GW : IG = 2 : 1$. There is an exact correspondence between the points O, G, T, and H on the Euler line, and the points I, G, J, and W on the second line. The Euler line is concerned with the circumscribed circles of the triangles ABC and LMN, and the second line is concerned with their inscribed circles. This is because I and J are the incentres of ABC and LMN, respectively. Also, W and G are the external and internal centres of similitude of the two incircles. This configuration is shown in Fig. 8.4.

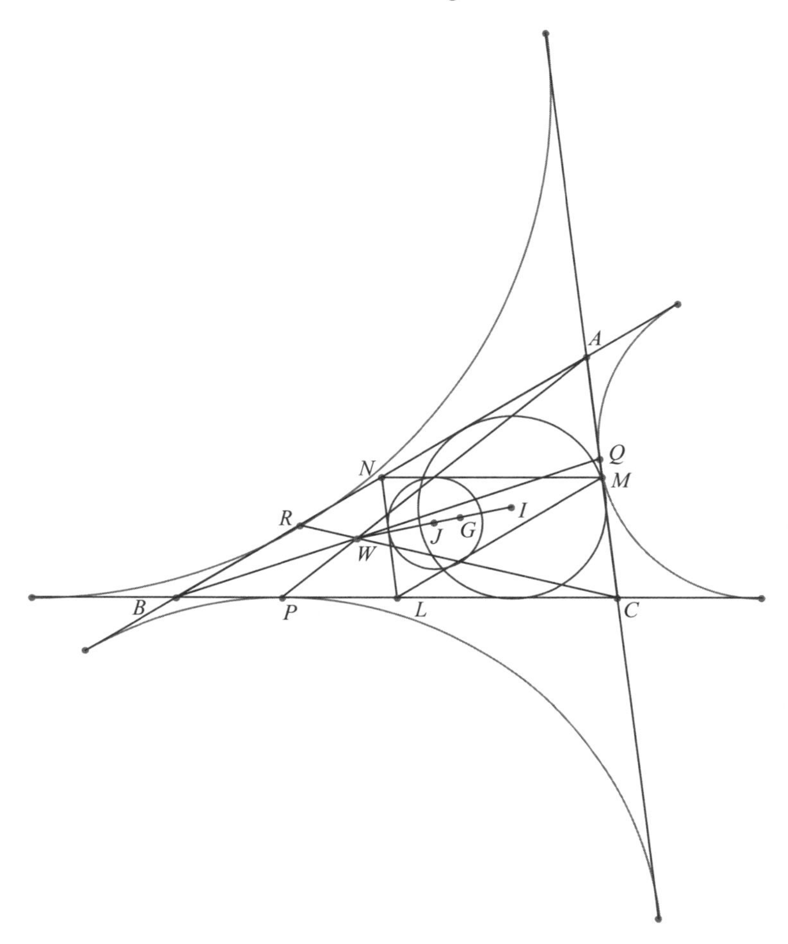

Fig. 8.4 The analogue of the Euler line for the two incircles.

Exercises 8.6

8.6.1 Show that the lines through L, M, and N parallel to the internal angle bisectors are concurrent at J, where J is the centre of mass of a uniform wire framework in the shape of the triangle ABC and L, M, and N are the midpoints of the sides.

8.6.2 Prove that the triangles IGH and JGO are similar, where the points are as defined in Section 8.6.

8.6.3 Prove, with the same notation as in Exercise 8.6.1, that J is the incentre of the triangle LMN.

8.6.4 Calculate IW^2, where I is the incentre and W is Nagel's point.

8.7 The triangle of excentres

We now consider the excircles and the triangle of excentres $I_1I_2I_3$. It turns out that, when the triangle ABC is integer-sided, all of the distances AI, BI, CI, AI_1, BI_2, CI_3, I_2I_3, I_3I_1, and I_1I_2 are rational whenever the internal bisectors are rational. It is shown in Section 2.8 how to arrange for this to happen, and the same method applies here.

The excircle opposite A touches the sides of ABC on the other side of BC from A. Its centre is denoted by I_1 and its radius by r_1. The excircles opposite B and C are similarly defined, having centres I_2 and I_3, and radii r_2 and r_3, respectively. See Section 2.6, where formulae for r_1, r_2, and r_3 are given, and it is shown that for a Heron triangle all of r, r_1, r_2, and r_3 are rational. The triangle $I_1I_2I_3$ is the triangle of excentres. The point I is the orthocentre of the triangle $I_1I_2I_3$, so ABC is its pedal triangle. The angles of $I_1I_2I_3$ are therefore $90° - \frac{1}{2}A$, $90° - \frac{1}{2}B$, and $90° - \frac{1}{2}C$, and its circumradius is $2R$. The circumcircle of the triangle ABC is the nine-point circle of the triangle $I_1I_2I_3$, and so passes through the midpoints of its sides.

We now give, without further proof, a catalogue of values for r, r_1, r_2, r_3, AI, AI_1, and I_2I_3 in terms of a, b, and c. The values are obtained from equations such as $\frac{1}{2}r_1(b+c-a) = [ABC]$, $AI = r \operatorname{cosec} \frac{1}{2}A$, and $AI_1 = r_1 \operatorname{cosec} \frac{1}{2}A$, and I_2I_3 is found from the areal metric, see Section A.9 of the appendix. We have

$$r = \frac{1}{2}\sqrt{\frac{(b+c-a)(c+a-b)(a+b-c)}{a+b+c}}, \tag{8.7.1}$$

$$r_1 = \frac{1}{2}\sqrt{\frac{(a+b+c)(c+a-b)(a+b-c)}{b+c-a}}, \tag{8.7.2}$$

with similar formulae for r_2 and r_3 by cyclic changes of letters. Also,

$$AI = \sqrt{\frac{bc(b+c-a)}{a+b+c}}, \tag{8.7.3}$$

with similar formulae for BI and CI by cyclic changes of letters, and

$$I_2I_3 = 2a\sqrt{\frac{bc}{(c+a-b)(a+b-c)}} = a \operatorname{cosec} \frac{1}{2}A = 4R\cos\frac{1}{2}A, \tag{8.7.4}$$

with similar expressions for I_3I_1 and I_1I_2 by cyclic changes of letters. It follows that, if $\sin\frac{1}{2}A$, $\sin\frac{1}{2}B$, $\sin\frac{1}{2}C$, $\cos\frac{1}{2}A$, $\cos\frac{1}{2}B$, and $\cos\frac{1}{2}C$ are rational, then AI, BI, CI, AI_1, BI_2, CI_3, I_2I_3, I_3I_1, and I_1I_2 are rational. See Section 2.8 for details of how to make the three internal angle bisectors rational. When these are made rational, then all of these other line segments have rational length, and by a suitable enlargement factor all can be made to have integer length.

Exercise 8.7

8.7.1 In the triangle in which $a = 154$, $b = 169$, and $c = 125$, find the lengths of AI_1, BI_2, CI_3, I_2I_3, I_3I_1, and I_1I_2, where I_1, I_2, and I_3 are the excentres opposite A, B, and C, respectively.

8.8 The lengths of OI and OH

We now turn our attention to the distances between the circumcentre O and the incentre I, and between O and the orthocentre H. We limit discussion to these cases, as they are representative of a vast number of distances in a triangle that might be studied, though in Section 8.9 we introduce a special set of distances that are rational for any Heron triangle. As we shall see below, the problem of making OI or OH rational for an integer-sided triangle is not at all easy and we have no solution to offer except in the case of an isosceles triangle.

The length of OI

In the triangle ABC let us draw the internal bisector of angle A. This passes through I and I_1, and we denote by P the point where it meets the circumcircle of the triangle ABC. The point I lies on the internal bisector of angle B and I_1 lies on the external bisector of angle B. It follows that $\angle IBI_1 = 90°$. Now $\angle I_1BP = \angle BI_1P = \frac{1}{2}C$, so $PB = PI_1$. Hence P is the centre of the circle IBI_1, so $IP = BP = \frac{1}{2}a \sec \frac{1}{2}A = 2R \sin \frac{1}{2}A$. Now, by the intersecting chord theorem, $AI \cdot IP = R^2 - OI^2$. However, $AI = r \operatorname{cosec} \frac{1}{2}A$ and $IP = BP = 2R \sin \frac{1}{2}A$, and hence $R^2 - OI^2 = 2Rr$. That is,

$$OI^2 = R^2 - 2Rr\,. \tag{8.8.1}$$

Now

$$R = \frac{abc}{4[ABC]} \quad \text{and} \quad r = \frac{2[ABC]}{a+b+c}\,. \tag{8.8.2}$$

Since

$$[ABC] = \frac{1}{4}\sqrt{(a+b+c)(b+c-a)(c+a-b)(a+b-c)} \tag{8.8.3}$$

and $\cos A = 1 - 2\sin^2 \frac{1}{2}A = (b^2 + c^2 - a^2)/2bc$, etc., it follows that

$$\sin\frac{1}{2}A \sin\frac{1}{2}B \sin\frac{1}{2}C = \frac{(b+c-a)(c+a-b)(a+b-c)}{8abc}\,,$$

and hence

$$r = 4R \sin\frac{1}{2}A \sin\frac{1}{2}B \sin\frac{1}{2}C\,. \tag{8.8.4}$$

Thus

$$
\begin{aligned}
OI^2 &= R^2 - 2Rr \\
&= R^2\left[1 - \frac{(b+c-a)(c+a-b)(a+b-c)}{abc}\right] \\
&= \frac{R^2(a^3+b^3+c^3-a^2b-b^2c-c^2a-b^2a-c^2b-a^2c+3abc)}{abc}.
\end{aligned}
\tag{8.8.5}
$$

As can be seen from eqns (8.8.2) and (8.8.3), R also depends on a, b, and c in quite a complicated way, so the general problem of finding integer a, b, and c to make OI rational is very difficult. When ABC is a Heron triangle we know that R is rational, so the problem when ABC is a Heron triangle is marginally less difficult.

However, there is one case that can be dealt with. The length of OI is rational when ABC is an isosceles Heron triangle. For example, when $a = 6$ and $b = c = 5$ we have $[ABC] = 12$, so $R = 25/8$. Then $OI = R|a-b|/b = 5/8$.

There are many cases when OI is a rational multiple of R, but unless R is rational, for example when ABC is a Heron triangle, these cases do not lead to a rational value of OI. For example, when $a = 52$, $b = 20$, and $c = 45$, then $OI = 13R/20$, but $[ABC] = (117/4)\sqrt{231}$, so R is irrational. As a second example, see Fig. 8.5, when $a = 7$, $b = 8$, and $c = 6$, then $OI = (1/4)R$, but $[ABC] = (21/4)\sqrt{15}$, and again R is irrational.

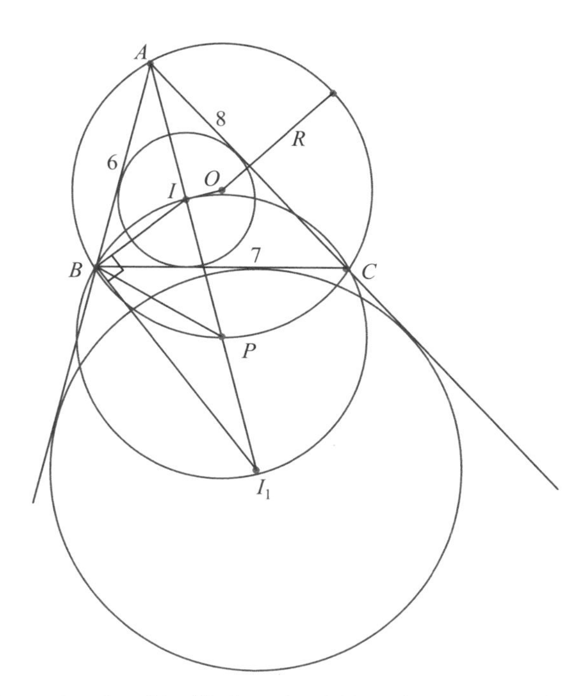

Fig. 8.5 In a triangle with side lengths 6, 7, and 8, we have $OI = (1/4)R$.

The length of OH

In the triangle ABC we have $\angle BOC = 2A$, etc., so

$$\begin{aligned} OH^2 &= (\boldsymbol{OA} + \boldsymbol{OB} + \boldsymbol{OC}) \cdot (\boldsymbol{OA} + \boldsymbol{OB} + \boldsymbol{OC}) \\ &= 3R^2 + 2R^2(\cos 2A + \cos 2B + \cos 2C) \\ &= R^2(1 - \cos A \cos B \cos C) \\ &= \frac{R^2(a^6 + b^6 + c^6 - a^4b^2 - b^4c^2 - c^4a^2 - b^4a^2 - c^4b^2 - a^4c^2 + 3a^2b^2c^2)}{a^2b^2c^2}. \end{aligned} \tag{8.8.6}$$

The general problem of finding an integer-sided triangle in which OH is rational is very difficult, but, as with OI, the length of OH is rational when ABC is an isosceles Heron triangle. For example, when $a = 6$ and $b = c = 5$ we have $R = 25/8$. Then $OH = R|a^2 - b^2|/b^2 = 11/8$. There are many cases when OH is a rational multiple of R, but again unless R is rational, for example when ABC is a Heron triangle, these cases do not lead to a rational value of OH. For example, when $a = 2$ and $b = c = 3$, which is isosceles but not a Heron triangle, we have $OH = 5R/9$, but $[ABC] = 2\sqrt{2}$, so R is not rational.

Exercises 8.8

8.8.1 Find OI in the isosceles Heron triangle with sides $a = 10$ and $b = c = 13$.

8.8.2 Find OH in the isosceles Heron triangle with sides $a = 10$ and $b = c = 13$.

8.9 Feuerbach's theorem

Theorem 8.9.1 (Feuerbach) *The nine-point circle touches the incircle and each of the excircles.* □

For a proof of this theorem we refer to Hahn (1994), where three proofs are given, two using complex numbers and one, the standard proof, using inversion. Feuerbach originally gave a trigonometrical proof, which is a straightforward but laborious method.

The theorem implies that $IT = \frac{1}{2}R - r$, $I_1T = \frac{1}{2}R + r_1$, $I_2T = \frac{1}{2}R + r_2$, and $I_3T = \frac{1}{2}R + r_3$. Hence all of these distances are rational for any Heron triangle.

Exercises 8.9

8.9.1 Find IT, I_1T, I_2T, and I_3T for the triangle with $a = 7$, $b = 15$, and $c = 20$.

8.9.2 Prove that $OW = R - 2r$, where W is Nagel's point.

9 Solids

In this chapter we consider, rather briefly, some problems associated with solids.

The first of these is the analogue in three dimensions of the problem of the Heron triangle in the plane. If we are given a tetrahedron with integer edge lengths, can these lengths be chosen so that the tetrahedron has integer volume? The answer, as one might guess, is that there are an infinite number, but none that are regular or semi-regular.

We then look at the various types of tetrahedra with integer edges and ask for which of them is it possible for the circumscribing sphere to have integer radius. In the course of Section 9.2 we prove two curious and little-known theorems.

Finally, in Section 9.3 we give a brief review of the regular solids and hypersolids in three and four dimensions, respectively. There is a long exercise on the thirteen semi-regular solids in three dimensions for the reader who is interested in their classification.

9.1 Tetrahedrons with integer edges and integer volume

We next consider various different types of tetrahedra, all of which have integer edge lengths, and we investigate for each type whether there are choices of edge lengths for which the volume is also an integer.

The tetrahedron $VABC$ in which $BC = a$, $CA = b$, $AB = c$, $VA = d$, $VB = e$, and $VC = f$ has volume $[VABC]$ given by

$$\begin{aligned}144[VABC]^2 = -a^2b^2c^2 + (a^2d^2 + e^2f^2)(b^2 + c^2 - a^2) \\ + (b^2e^2 + f^2d^2)(c^2 + a^2 - b^2) + (c^2f^2 + d^2e^2)(a^2 + b^2 - c^2) \\ - a^2d^4 - b^2e^4 - c^2f^4 .\end{aligned} \tag{9.1.1}$$

Methods for deriving this formula, first obtained by Euler, are given in Salmon (1912).

The regular tetrahedron

A *regular tetrahedron* is one in which all the edges are equal. If all of the edges are equal to a, then from eqn (9.1.1) we have

$$[VABC]^2 = \frac{a^6}{72}, \tag{9.1.2}$$

and so a and $[VABC]$ cannot both have integer values.

The semi-regular tetrahedron

A *semi-regular tetrahedron* is one in which $a = b = c$ and $d = e = f$, and then from eqn (9.1.1) we have

$$[VABC]^2 = \frac{a^4(3d^2 - a^2)}{144}. \tag{9.1.3}$$

If a and d are integers then, as can be seen by working modulo 4, there is no integer n such that $a^2 + n^2 = 3d^2$, so, once more, if the sides are integers then $[VABC]$ cannot have an integer value.

The balloon tetrahedron

A *balloon tetrahedron* $PQRS$ is defined to be one in which the vertices P, Q, R, and S are the centres of four mutually touching spheres, of radii p, q, r, and s, respectively. A balloon tetrahedron is a type of tetrahedron that depends on four, rather than six, parameters.

Putting $a = r + s$, $b = s + q$, $c = q + r$, $d = p + q$, $e = p + r$, and $f = p + s$ in eqn (9.1.1), we obtain

$$144[PQRS]^2 = 16p^2q^2r^2s^2 \left[2\left(\frac{1}{pq} + \frac{1}{pr} + \frac{1}{ps} + \frac{1}{qr} + \frac{1}{qs} + \frac{1}{sr}\right) - \left(\frac{1}{p^2} + \frac{1}{q^2} + \frac{1}{r^2} + \frac{1}{s^2}\right)\right]. \tag{9.1.4}$$

Writing $p = 1/w$, $q = 1/x$, $r = 1/y$, and $s = 1/z$, where w, x, y, and z are the curvatures of the spheres, we obtain

$$3[PQRS]wxyz = \sqrt{2(wx + wy + wz + xy + xz + yz) - (w^2 + x^2 + y^2 + z^2)}. \tag{9.1.5}$$

In Section 7.3 we have seen how to make the right-hand side equal to an integer divided by $\sqrt{3}$. If we add a fifth sphere, inside the tetrahedron, of radius t, curvature $v = 1/t$, and centre T, so that it touches the other four spheres externally, then the equation

$$[TPQR] + [TPQS] + [TPRS] + [TQRS] = [PQRS] \tag{9.1.6}$$

becomes

$$v[w, x, y, z] = w[v, x, y, z] + x[v, w, y, z] + y[v, w, x, z] + z[v, w, x, y], \tag{9.1.7}$$

so that w, x, y, z, and v form a touching set (since we can introduce $\sqrt{3}$ in each term of eqn (9.1.6)), see Exercise 7.3.5. However, in order to make $[PQRS]$ integral we have a slightly different expression to deal with (without the factor 3 under the square root). However, it is not difficult to make the expression

$$2(wx + wy + wz + xy + xz + yz) - (w^2 + x^2 + y^2 + z^2) = \{w, x, y, z\}^2 \tag{9.1.8}$$

a perfect square.

Theorem 9.1.1 *If we choose positive integers B, C, and D such that*

$$4(BC + CD + DB) = h^2 + k^2\,, \tag{9.1.9}$$

where h and k are positive integers, and if $A = B+C+D+k$, then $\{A, B, C, D\} = h$, where $\{A, B, C, D\}$ is given by eqn (9.1.8).

Proof We have

$$\begin{aligned}\{A, B, C, D\}^2 &= 2(BC + CD + DB) + 2(B + C + D)A \\ &\qquad - (A^2 + B^2 + C^2 + D^2) \\ &= 2(BC + CD + DB) + 2(B + C + D)(B + C + D + k) \\ &\qquad - \left[(B + C + D + k)^2 + B^2 + C^2 + D^2\right] \\ &= 4(BC + CD + DB) - k^2 \\ &= h^2\,,\end{aligned}$$

where terms involving $k(B + C + D)$, B^2, C^2, and D^2 all cancel. □

For example, when $B = 1$, $C = 2$, and $D = 5$ then $4(BC + CD + DB) = 68 = 8^2 + 2^2$. So, if we choose $A = 10$ then we obtain $\{10, 1, 2, 5\} = 8$. The geometrical significance is that with $w = 10$, $x = 1$, $y = 2$, and $z = 5$ we obtain $210[PQRS] = 8$ or $[PQRS] = 4/105$. This corresponds to $p = 1/10$, $q = 1$, $r = 1/2$, and $s = 1/5$. Multiplying up by 10, we obtain $p = 1$, $q = 10$, $r = 5$, $s = 2$, and $[PQRS] = 800/21$. So the tetrahedron with sides 11, 6, 3, 15, 12, and 7 has rational volume $800/21$, see Fig. 9.1. With the edges a, b, c, d, e, and f given in terms of as many as four parameters, there are an infinite number of solutions of this type.

The isosceles tetrahedron

An *isosceles tetrahedron* is a tetrahedron in which $a = d$, $b = e$, and $c = f$, and eqn (9.1.1) now gives

$$72[VABC]^2 = (b^2 + c^2 - a^2)(c^2 + a^2 - b^2)(a^2 + b^2 - c^2)\,. \tag{9.1.10}$$

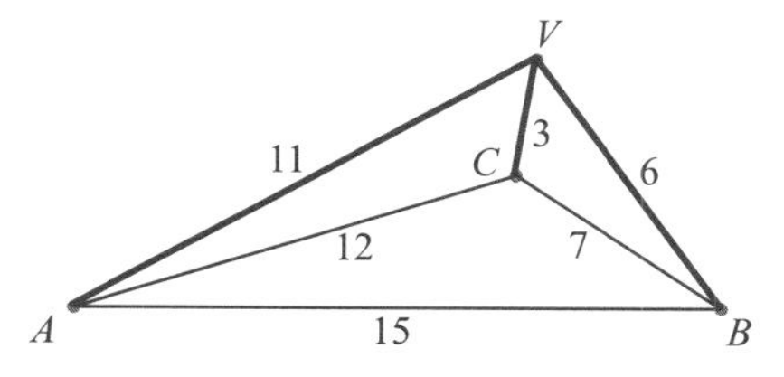

Fig. 9.1 A balloon tetrahedron with integer edges and rational volume $800/21$.

It is possible to find integers a, b, and c so that the right-hand side is twice a perfect square. This can be done, for example, by using the results of Section 1.4, though perhaps there are easier ways. We look for x, y, and z such that

$$\begin{aligned} b^2 + c^2 - a^2 &= 2y^2z^2\,, \\ c^2 + a^2 - b^2 &= 2z^2x^2\,, \\ a^2 + b^2 - c^2 &= 2x^2y^2\,. \end{aligned} \tag{9.1.11}$$

Then

$$\begin{aligned} a^2 &= x^2(y^2 + z^2)\,, \\ b^2 &= y^2(z^2 + x^2)\,, \\ c^2 &= z^2(x^2 + y^2)\,. \end{aligned} \tag{9.1.12}$$

If x, y, and z are the sides of a rectangular box with integer face diagonals then we have a solution. For example, with $x = 44$, $y = 117$, and $z = 240$ we have $a = 44 \times 267 = 11\,748$, $b = 117 \times 244 = 28\,548$, and $c = 240 \times 125 = 30\,000$. Then $72[VABC]^2 = 8x^4y^4z^4$ and $[VABC] = x^2y^2z^2/3 = 508\,836\,556\,800$. The sides can all be divided by 4 to give the solution $a = 2937$, $b = 7137$, $c = 7500$, and $[VABC] = 7\,950\,571\,200$, see Fig. 9.2.

The orthogonal tetrahedron

An *orthogonal tetrahedron* is a tetrahedron in which $a^2+d^2 = b^2+e^2 = c^2+f^2 = k^2$, for some constant k. From eqn (9.1.1) we have

$$\begin{aligned} [VABC]^2 &= \frac{1}{144}\left[k^2(2b^2c^2 + 2c^2a^2 + 2a^2b^2 - a^4 - b^4 - c^4) - 4a^2b^2c^2\right] \\ &= \frac{1}{144}\left\{(16k^2 - 64R^2)[ABC]^2\right\} \\ &= \frac{1}{9}(k^2 - 4R^2)[ABC]^2\,, \end{aligned} \tag{9.1.13}$$

where R is the circumradius of the triangle ABC. Since this is positive, it follows that for an orthogonal tetrahedron $k > 2R$.

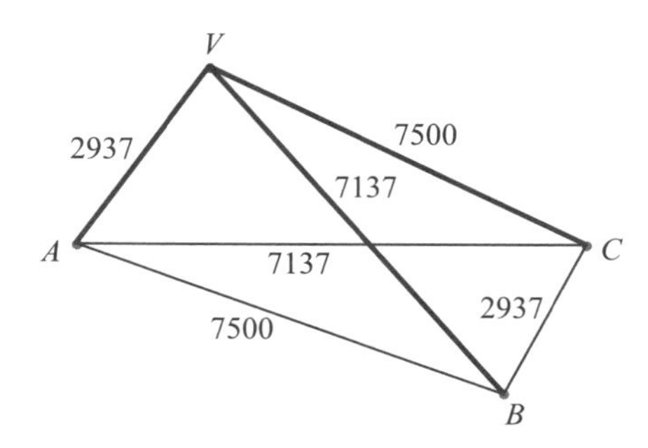

Fig. 9.2 An isosceles tetrahedron with integer edges and integer volume 7 950 571 200.

It is possible to find an orthogonal tetrahedron with integer sides and integer volume, by once again using the data from the rectangular box in Section 1.4. For example, if we take $d = 44$, $e = 117$, $f = 240$, $a = 267$, $b = 244$, and $c = 125$, then $a^2 + d^2 = b^2 + e^2 = c^2 + f^2 = k^2 = 73\,225$, and, though k is irrational, we still have a tetrahedron of integer volume. This is because VA, VB, and VC form a right triad of axes, and so $[VABC] = (1/6) \times 44 \times 117 \times 240 = 205\,920$.

Exercises 9.1

9.1.1 Choose $B = 2$, $C = 3$, and $D = 4$ in Theorem 9.1.1 and hence determine a balloon tetrahedron with integer edges and rational volume.

9.1.2 Find the volume of a tetrahedron with edge lengths $a = 11$, $b = 10$, $c = 9$, $d = 8$, $e = 7$, and $f = 6$.

9.2 The circumradius of a tetrahedron

Tetrahedral co-ordinates

One of the main algebraic resources for solving problems about the tetrahedron is the use of tetrahedral (or volumetric) co-ordinates. These are the analogue in three dimensions of areal (or barycentric) co-ordinates in two dimensions, and we now give a brief introduction to the use of these co-ordinates. A full account of areal co-ordinates is given in the appendix.

In tetrahedral co-ordinates a point P is given four co-ordinates (w, x, y, z), where

$$w = \frac{[PABC]}{[VABC]}, \quad x = \frac{[VPBC]}{[VABC]}, \quad y = \frac{[VAPC]}{[VABC]}, \quad z = \frac{[VABP]}{[VABC]} \qquad (9.2.1)$$

and all co-ordinates are signed. For example, w is signed so that it is positive if P is on the same side of the plane ABC as V, and negative if P is on the opposite side of ABC as V.

By definition, for any point P, $w + x + y + z = 1$. However, co-ordinates are often given in unnormalised form to avoid cumbersome expressions, and because, for many purposes, only the ratios of the co-ordinates are needed.

If we have $P(k, l, m, n)$ and $Q(w, x, y, z)$ then $\boldsymbol{PQ} = (w - k, x - l, y - m, z - n)$ is a vector displacement and provides both the direction of the displacement and its magnitude. If $\mathbf{d}\boldsymbol{s} = (\mathrm{d}w, \mathrm{d}x, \mathrm{d}y, \mathrm{d}z)$ is a displacement then $\mathrm{d}w + \mathrm{d}x + \mathrm{d}y + \mathrm{d}z = 0$ and the square of its magnitude is given by the formula

$$(\mathrm{d}s)^2 = -a^2\,\mathrm{d}y\,\mathrm{d}z - b^2\,\mathrm{d}z\,\mathrm{d}x - c^2\,\mathrm{d}x\,\mathrm{d}y - d^2\,\mathrm{d}w\,\mathrm{d}x - e^2\,\mathrm{d}w\,\mathrm{d}y - f^2\,\mathrm{d}w\,\mathrm{d}z\,. \quad (9.2.2)$$

In eqn (9.2.2) the edge lengths a, b, c, d, e, and f are as defined in Section 9.1. Note that eqn (9.2.2) is valid, not only for an infinitesimal arc length, but also when $\mathbf{d}\boldsymbol{s}$ is a finite straight-line displacement.

Given P and Q with co-ordinates as above, then the point R with co-ordinates $[p(k,l,m,n)+q(w,x,y,z)]/(p+q)$ lies on the line PQ and is such that $PR/RP = q/p$, where the line segments are signed. If P, Q, and R are any three points then the equation of the plane containing them is the 4×4 determinant with (w,x,y,z) as its first row and the co-ordinates of P, Q, and R as the second, third, and fourth rows, respectively. Four points P, Q, R, and S are therefore coplanar if the determinant with their co-ordinates as rows vanishes.

We do not give proofs of any of the above results, but rely on the natural generalisation of the theory of areal co-ordinates, given in the appendix.

The circumradius of the general tetrahedron

In this section we use the formula (9.1.1) for the volume of $VABC$, given at the beginning of Section 9.1. We also use the equation of the circumsphere in tetrahedral co-ordinates, which is

$$a^2yz + b^2zx + c^2xy + d^2wx + e^2wy + f^2wz = 0\,. \tag{9.2.3}$$

Consider the diameter VW of the circumsphere passing through its midpoint O. We have $VW^2 = 4R^2$, where R is the circumradius. Now V has co-ordinates $(1,0,0,0)$ and suppose that W has co-ordinates $(2w_0, 2x, 2y, 2z)$, so that O has co-ordinates $(\frac{1}{2}+w_0, x, y, z)$ and $w_0+x+y+z=\frac{1}{2}$. We next determine VW^2 using the metric (9.2.2). Since $\boldsymbol{VW} = (2w_0-1, 2x, 2y, 2z)$, we have

$$\begin{aligned} 4R^2 &= VW^2 \\ &= -a^2(4yz) - b^2(4zx) - c^2(4xy) - d^2(2x)(2w_0-1) \\ &\qquad\qquad - e^2(2y)(2w_0-1) - f^2(2z)(2w_0-1) \\ &= 2xd^2 + 2ye^2 + 2zf^2\,, \end{aligned} \tag{9.2.4}$$

where we have used the fact that W lies on the circumsphere with equation (9.2.3). Writing $4R^2 = 2s$, we have

$$d^2x + e^2y + f^2z - s = 0\,, \tag{9.2.5}$$

where x, y, and z are three of the co-ordinates of the circumcentre. It follows by similar arguments that, if w is the other co-ordinate, then

$$c^2y + b^2z + d^2w - s = 0\,, \tag{9.2.6}$$

$$c^2x + a^2z + e^2w - s = 0\,, \tag{9.2.7}$$

$$b^2x + a^2y + f^2w - s = 0\,. \tag{9.2.8}$$

Equations (9.2.5)–(9.2.8), together with the equation $x+y+z+w=1$, may be solved to determine $s = 2R^2$ (and w, x, y, and z if required). This calculation has been checked using DERIVE, as developed by Texas Instruments Inc. and the authors of the software at Soft Warehouse Inc., Honolulu, Hawaii, and, using the formula for $[VABC]$, we find the result to be

$$576R^2[VABC]^2 = (ad+be+cf)(-ad+be+cf)(ad-be+cf)(ad+be-cf)\,. \tag{9.2.9}$$

These results may be summarised in the following two remarkable and curious theorems.

Theorem 9.2.1 *If $VABC$ is a tetrahedron with $BC=a$, $CA=b$, $AB=c$, $VA=d$, $VB=e$, $VC=f$, volume $[VABC]$, and circumradius R, then $6R[VABC]=\Delta$, where Δ is the area of a triangle with sides ad, be, and cf (regarded as lengths with these magnitudes).* □

Theorem 9.2.2 *If $VABC$ and $KLMN$ are two tetrahedra with sides $BC=KL$, $CA=KM$, $AB=KN$, $VA=MN$, $VB=NL$, and $VC=LM$, and if R and S are their respective circumradii, then $R[VABC]=S[KLMN]$.* □

The complexity of the formula for $[VABC]$ (see Section 9.1) makes it impractical to treat the problem of finding general integer-sided tetrahedra with rational circumradius, but there are solutions in certain special cases.

The circumradius of the isosceles tetrahedron

In an isosceles tetrahedron, in which $d=a$, $e=b$, and $f=c$, we have

$$576R^2[VABC]^2 = (a^2+b^2+c^2)(b^2+c^2-a^2)(c^2+a^2-b^2)(a^2+b^2-c^2)\,, \tag{9.2.10}$$

but from Section 9.1 we know that

$$72[VABC]^2 = (b^2+c^2-a^2)(c^2+a^2-b^2)(a^2+b^2-c^2)\,,$$

and hence

$$R^2 = \frac{1}{8}(a^2+b^2+c^2)\,. \tag{9.2.11}$$

The formula for R^2 in this case may be more easily obtained by using the fact that the circumcentre O of an isosceles tetrahedron coincides with its centroid G at the point with co-ordinates $(1/4, 1/4, 1/4, 1/4)$.

In order to find a three-parameter system for integer values of R in terms of integer values of a, b, and c, we follow the usual procedure of finding a non-singular linear change of variables, so that eqn (9.2.11) becomes linear in each variable. One such change of variables is

$$\begin{aligned} a &= 2v+2w+8x\,,\\ b &= 2u+2w-2x\,,\\ c &= 2u+2v-2x\,,\\ R &= u+v+w+3x\,. \end{aligned} \tag{9.2.12}$$

Equation (9.2.11) now gives $u = (-vw - 3vx - 3wx)/(v + w + 8x)$. Substituting back the value of u into eqns (9.2.12) and multiplying up by $v + w + 8x$, we obtain

$$\begin{aligned} a &= 2(v + w + 4x)(v + w + 8x)\,, \\ b &= 2(w^2 + 4wx - 8x^2 - 4vx)\,, \\ c &= 2(v^2 + 4vx - 8x^2 - 4wx)\,, \\ R &= v^2 + w^2 + vw + 8vx + 8wx + 24x^2\,. \end{aligned} \tag{9.2.13}$$

It is possible that common factors may have to be extracted from a, b, c, and R to obtain a primitive set of values with highest common factor 1. Also, some of the variables a, b, c, and R may have to be changed in sign, but that is no problem since each variable appears only as a squared quantity.

As an example, $x = 1$, $v = -5$, and $w = -7$ gives $a = 64$, $b = 66$, $c = 50$, and $R = 37$. As a second example, $x = 1$, $v = -4$, and $w = -7$ gives $a = 42$, $b = 58$, $c = 40$, and $R = 29$. The second example is shown in Fig. 9.3.

The circumradius of a regular tetrahedron

The formula (9.2.9) gives $R = (3/4)a\sqrt{2}$, since $[VABC] = a^6/72$. It follows that R is never an integer when a is an integer.

The circumradius of a semi-regular tetrahedron

The formula (9.2.9) gives

$$576R^2[VABC]^2 = 3a^4d^4\,, \tag{9.2.14}$$

where a is the side of the base and d is the length of the sloping edges. By eqn (9.1.3), $[VABC]^2 = a^4(3d^2 - a^2)/144$, and hence

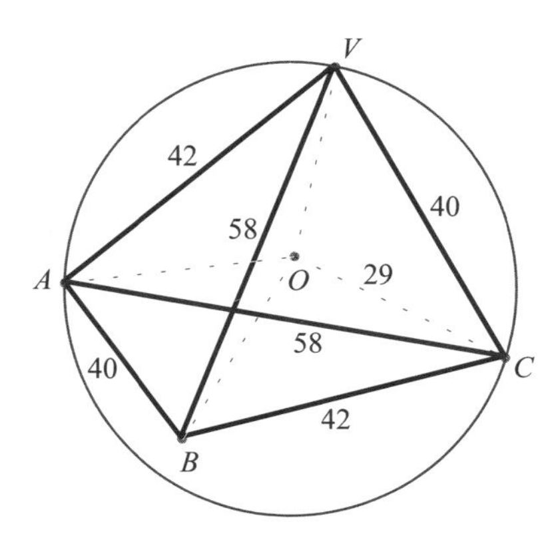

Fig. 9.3 An isosceles tetrahedron with integer edges and integer circumradius.

$$R^2 = \frac{3d^4}{4(3d^2 - a^2)}. \quad (9.2.15)$$

Although $[VABC]$ is not rational for integral a and d, there is a two-parameter set of values of a and d for which R is rational. All we have to do is to solve the equation

$$3d^2 - a^2 = 3x^2 \quad (9.2.16)$$

in integers, and then $R = d^2/2x$. The general two-parameter solution of eqn (9.2.16) is

$$\begin{aligned} d &= 2(v^2 + vw + w^2), \\ a &= 3w(2v + w), \\ x &= |w^2 - 2vw - 2v^2|, \end{aligned} \quad (9.2.17)$$

where common factors may be introduced or extracted. For example, when $v = 1$ and $w = 1$ we obtain $d = 2$, $x = 1$, $a = 3$, and $R = 2$. As a second example, when $v = 1$ and $w = 3$ we obtain $d = 26$, $x = 1$, $a = 45$, and $R = 338$.

Exercises 9.2

9.2.1 Find the sides and circumradius of the isosceles tetrahedron that results from the parameters $x = 2$, $v = -7$, and $w = 5$ in eqns (9.2.12).

9.2.2 Find the sides and circumradius of the semi-regular tetrahedron that results from the parameters $v = 3$ and $w = 2$ in eqns (9.2.17).

9.3 The five regular solids and six regular hypersolids

In this section we give a brief survey of the five regular solids in three dimensions and the six regular hypersolids in four dimensions. In Exercises 9.3 the reader is encouraged to consider the thirteen semi-regular solids.

The five regular solids

Suppose that a regular solid has F regular n-gons and that at every vertex of such an n-gon m edges (faces) come together. The angles of a regular polygon with $n = 3$, 4, 5, and 6 are $60°$, $90°$, $108°$, and $120°$, respectively. The sum of the angles at any vertex must be less than $360°$, and there must be at least three polygons at each vertex. Hence the only possibilities are $n = 3$ and $m = 3$, 4, or 5, or $n = 4$ and $m = 3$, or $n = 5$ and $m = 3$. The number of vertices is $V = nF/m$ and the number of edges is $E = nF/2$. Now Euler's relation is $F + V = E + 2$, so we have $F + nF/m = nF/2 + 2$, and $2/n + 2/m = 1 + 4/nF$. All five possibilities give integer values for F, V, and E, and they are shown in Table 9.1.

Note the duality between the octahedron and the cube, and between the icosahedron and the dodecahedron. These solids are sometimes referred to as the Platonic solids.

Table 9.1 The five regular solids.

	n	m	F	V	E
Tetrahedron	3	3	4	4	6
Octahedron	3	4	8	6	12
Icosahedron	3	5	20	12	30
Cube	4	3	6	8	12
Dodecahedron	5	3	12	20	30

The six regular hypersolids

If you start with a line, which has one edge and two vertices, and move it at right angles to its direction the correct distance, then one gets a square. The two vertices are replicated by two more at the end position of the movement, making four vertices for the square. The two vertices, as they are drawn along, produce two lines, and there are the starting and finishing positions of the original line, making four edges for the square. The line, as it moves, creates the face of the square. In an obvious notation, where the subscript denotes the dimension, $V_2 = 2V_1 = 4$, $E_2 = V_1 + 2E_1 = 4$, and $F_2 = E_1 = 1$.

If you now take the square and move it the correct distance perpendicular to its face, then you create the cube, and by a similar argument to before $V_3 = 2V_2 = 8$, $E_3 = V_2 + 2E_2 = 12$, $F_3 = E_2 + 2F_2 = 6$, and $C_3 = F_2 = 1$ (cube). It is clear that this process may be continued into a fourth and higher dimension, so that if we move the cube perpendicular to itself within the fourth dimension then we obtain a hypercube with $V_4 = 2V_3 = 16$, $E_4 = V_3 + 2E_3 = 32$, $F_4 = E_3 + 2F_3 = 24$, $C_4 = F_3 + 2C_3 = 8$, and $H_4 = C_3 = 1$ (hypercube). So a hypercube has 8 cubical cells, 24 faces, 32 edges, and 16 vertices.

The relation in four dimensions corresponding to Euler's relation is $V+F = E+C$, where C is the number of solid cells bounding the hypersolid. There are six convex regular hypersolids, including the hypercube. They are summarised in Table 9.2.

The simplex is the generalisation of the triangle and the tetrahedron; its cells are tetrahedral, and its faces are triangular. It is self-dual. The 16-cell hypersolid is the

Table 9.2 The six regular hypersolids.

	V	F	E	C
Simplex	5	10	10	5
Hypercube	16	24	32	8
16-cell hypersolid	8	32	24	16
24-cell hypersolid	24	96	96	24
120-cell hypersolid	600	720	1200	120
600-cell hypersolid	120	1200	720	600

dual of the hypercube. The 24-cell hypersolid is self-dual, and the other pair are duals of one another. If one moves into yet higher dimensions then only the generalisations of the simplex, the hypercube, and its dual exist, and they are the only three regular convex polytopes.

Exercises 9.3

9.3.1 There are thirteen Archimedean convex polyhedra, whose faces are all regular polygons, but of two or more kinds, and whose vertices are identical. They are listed below together with the type and number of polygons at each vertex. Complete the classification by enumerating in each case the values of F, V, and E.

(i) Truncated tetrahedron: 2 hexagons and 1 triangle.

(ii) Truncated cube: 2 octagons and 1 triangle.

(iii) Cuboctahedron: 2 squares and 2 triangles.

(iv) Truncated octahedron: 2 hexagons and 1 square.

(v) Small rhombicuboctahedron: 3 squares and 1 triangle.

(vi) Great rhombicuboctahedron: 1 octahedron, 1 hexagon, and 1 square.

(vii) Snub cube: 1 square and 4 triangles.

(viii) Icosidodecahedron: 2 pentagons and 2 triangles.

(ix) Truncated dodecahedron: 2 decagons and 1 triangle.

(x) Truncated icosahedron: 2 hexagons and 1 pentagon.

(xi) Small rhombicosidodecahedron: 1 pentagon, 2 squares, and 1 triangle.

(xii) Great rhombicosidodecahedron: 1 decagon, 1 hexagon, and 1 square.

(xiii) Snub dodecahedron: 1 pentagon and 4 triangles.

9.3.2 Find the number of vertices, edges, faces, three-dimensional cells, and four-dimensional hypersolids in a five-dimensional simplex, and in a five-dimensional cubic polytope.

10 Circles and conics

This chapter deals with some infinite sets of circles and conics. The infinite sequence of circles featured in Section 10.1 all have the same unit radius, and have intersection properties that may be seen in Fig. 10.1, where the start of such a sequence is shown. For further reading on infinite sets of circles see the book Hahn (1994).

The infinite set of conics are those that circumscribe a triangle. In Section 10.2 we show how to generalise the Simson line property, normally associated with the circumcircle. It turns out that there is a circumscribing conic associated with each point Q in the plane of a triangle, not lying on its sides, and that each point on the conic has its own associated line. Given the point Q, if G is the centroid, then the point Q_0 such that $\boldsymbol{GQ} = 2\boldsymbol{Q_0G}$ is the centre of the circumscribing conic, so that, as with points on the Euler line, the ratio $2 : 1$ appears. When the circumscribing conic is the circumcircle, the point Q is the orthocentre H, and the point Q_0 is the circumcentre O.

A more detailed account of these conics and the lines associated with points on them may be found in Bradley and Bradley (1996) and an extension of the work may be found in Bradley (2002). Just as there is the nine-point circle that has a radius one-half that of the circumcircle, there is associated with each of these conics another conic, passing through the midpoints of the sides of the triangle and the midpoints of AQ, BQ, and CQ, which is called the nine-point conic. This is described in Section 10.3.

10.1 Sequences of intersecting circles of unit radius

Let (z_k) be any sequence of complex numbers satisfying $|z_k| = 1$ for all k. Consider the circles C_k with equation $|z - z_k| = 1$, centre z_k, and radius 1. Since $|z_k| = 1$, it follows that all of these circles pass through O, the origin of the Argand diagram. Take any pair of these circles, say C_1 and C_2. Since $|z_1 + z_2 - z_1| = 1$ and $|z_1 + z_2 - z_2| = 1$, C_1 and C_2 meet at the point $z_1 + z_2$, as well as at O. Hence all of these circles meet in pairs at points $z_j + z_k$ $(j \neq k)$.

Consider now any three of the circles, say C_1, C_2, and C_3. They all pass through O and their other points of intersection are $z_2 + z_3$, $z_3 + z_1$, and $z_1 + z_2$. The circle C_{123}, centre $z_1 + z_2 + z_3$, and radius 1, passes through all three points of intersection, since $|z_1| = |z_2| = |z_3| = 1$. There is an infinite sequence of such circles C_{jkl} $(j \neq k$, $k \neq l$, $l \neq j)$. Consider now the four circles C_{234}, C_{134}, C_{124}, and C_{123} with centres $z_2 + z_3 + z_4$, $z_1 + z_3 + z_4$, $z_1 + z_2 + z_4$, and $z_1 + z_2 + z_3$, respectively. The circle C_{1234}, centre $z_1 + z_2 + z_3 + z_4$, and radius 1, evidently passes through the centres of the four circles C_{234}, C_{134}, C_{124}, and C_{123}. There is an infinite sequence of such circles C_{jklm}, no two of the subscripts being equal. Now consider the five circles C_{2345}, C_{1345}, C_{1245}, C_{1235}, and C_{1234}. Then there is a circle C_{12345} that passes through the centres of the five circles. The result just goes on and on, with sequences of circles passing through the centres of the previous sequence. Figure 10.1 shows the start of the sequence.

There are other chains of theorems involving sequences of circles, which depend for their proof on the pivot theorem and the six-circle theorem, see Section 8.4 and Exercises 10.1.

Exercises 10.1

10.1.1 Four lines in general position in a plane determine four triangles, when the lines are taken in threes. Show, using the pivot theorem (see Section 8.4), that the circumcircles of the four triangles pass through a point, often called *the Clifford point* of the four lines.

10.1.2 Consider five lines in a plane, no two of which are parallel and no three of which are concurrent. From Exercise 10.1.1, if we take any four of them it leads to four circles passing through a point. In all, there are five such Clifford points. Use the six-circle theorem to show that these five points lie on a circle, often called the *Clifford circle* of the five lines. What happens if you start with six lines in general position?

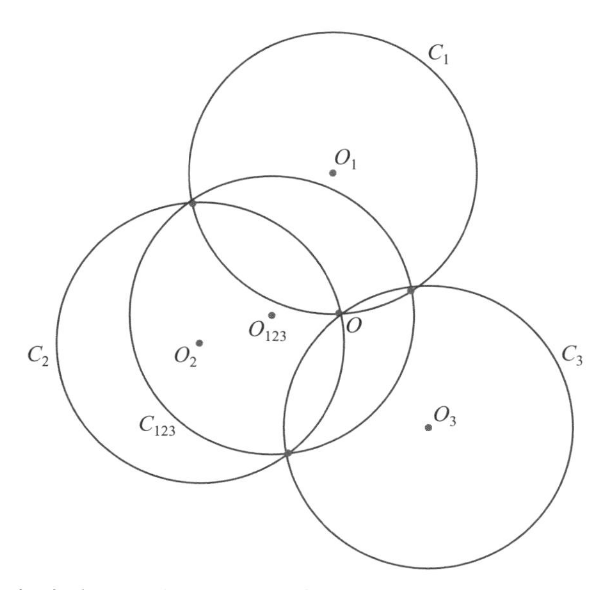

Fig. 10.1 Three unit circles passing through O and the unit circle through their other points of intersection.

10.2 Simson lines and Simson conics

Theorem 10.2.1 (The Simson line) *If ABC is a triangle and P is a point on its circumcircle, other than at the vertices, and if the feet of the perpendiculars from P onto the sides BC, CA, and AB are denoted by L, M, and N, respectively, then L, M, and N are collinear.* □

Theorem 10.2.1 is proved in any good textbook on geometry, for example, Durell (1946). In this section we show how to generalise this result.

Firstly, we provide some definitions. Given a triangle ABC and points P and Q in the plane of the triangle, let lines through P parallel to AQ, BQ, and CQ meet the sides BC, CA, and AB, respectively, at L, M, and N. We shall establish shortly that L, M, and N are collinear if and only if P lies on a certain circumscribing conic of ABC. This conic is called the *Simson conic* of Q with respect to ABC, and LMN is called the *Simson line* of P with respect to Q. When Q is at H, the orthocentre, the Simson conic is the circumscribing circle, centre O, and the perpendiculars from P onto the sides are lines through P parallel to AH, BH, and CH. Furthermore, as is shown in Section 8.6, $GH : OG = 2 : 1$. It is these properties that generalise as follows.

Theorem 10.2.2 *Let Q be a general point in the plane of the triangle ABC. Then the Simson conic of Q is the (unique) conic passing through A, B, and C with centre Q_0, where Q_0 lies on GQ, G is the centroid of ABC, and Q_0 is such that $\boldsymbol{GQ} : \boldsymbol{Q_0G} = 2 : 1$. This conic has the property that, if P lies on it and the lines parallel to AQ, BQ, and CQ are drawn through P to meet the sides BC, CA, and AB at L, M, and N, respectively, then L, M, and N are collinear.*

We also prove the following theorem.

Theorem 10.2.3 *If P lies on the Simson conic of Q, then Q lies on the Simson conic of P. The Simson line of P with respect to Q, the Simson line of Q with respect to P, and PQ are concurrent at the midpoint of PQ.*

It is also well documented in the literature that, if P lies on the circumcircle, then the Simson line of P with respect to H bisects PH (and therefore lies on the nine-point circle). Theorems 10.2.2 and 10.2.3 generalise what is already known about the original Simson line. (Note that the name of Simson is another misattribution. The result should be attributed to the Scottish mathematician Wallace.)

Proof of Theorems 10.2.2 and 10.2.3 We use areal co-ordinates. Let Q have co-ordinates (u, v, w) and let P be the general point with co-ordinates (x, y, z). Then $\boldsymbol{AQ} = (u-1, v, w) = (-(v+w), v, w)$. Points on the line through P parallel to $\boldsymbol{AQ}$ have co-ordinates of the form $(x - k(v+w), y+kv, z+kw)$. The point L where this line meets BC has unnormalised co-ordinates $(0, y(v+w)+xv, z(v+w)+xw)$. Similarly, M and N have unnormalised co-ordinates $(x(w+u)+yu, 0, z(w+u)+yw)$ and $(x(u+v)+zu, y(u+v)+zv, 0)$, respectively. The condition for LMN to be a straight line is that the determinant with these co-ordinates as its three rows should vanish. Using $x + y + z = u + v + w = 1$ this condition reduces to

$$u(v + w)yz + v(w + u)zx + w(u + v)xy = 0\,, \tag{10.2.1}$$

showing that the locus of P is a conic through A, B, and C. This is the Simson conic of Q with respect to ABC. The centre of this conic is the point Q_0 with co-ordinates $\frac{1}{2}(1 - u, 1 - v, 1 - w)$. The reason for this is that the points A_0, B_0, and C_0 with co-ordinates $(-u, 1 - v, 1 - w)$, $(1 - u, -v, 1 - w)$, and $(1 - u, 1 - v, -w)$, respectively, lie on the conic with equation (10.2.1), and Q_0 is the midpoint of AA_0, BB_0, and CC_0. Also, it is easily checked, since G has co-ordinates $(1/3, 1/3, 1/3)$, that $\boldsymbol{GQ} = 2\boldsymbol{Q}_0\boldsymbol{G}$. This establishes Theorem 10.2.2, see Fig. 10.2. Next, since eqn (10.2.1) may be rewritten as

$$x(y + z)vw + y(z + x)wu + z(x + y)uv = 0\,, \tag{10.2.2}$$

it follows that, if P lies on the Simson conic of Q, then Q lies on the Simson conic of P. Finally, if one adds the co-ordinates of L, M, and N together then one obtains $(u + x, v + y, w + z)$, and this means that the point with co-ordinates $\frac{1}{2}(u+x, v+y, w+z)$ lies on the line LMN. That is, the Simson line of P with respect to Q, and the Simson line of Q with respect to P both pass through the midpoint of PQ. This establishes Theorem 10.2.3. □

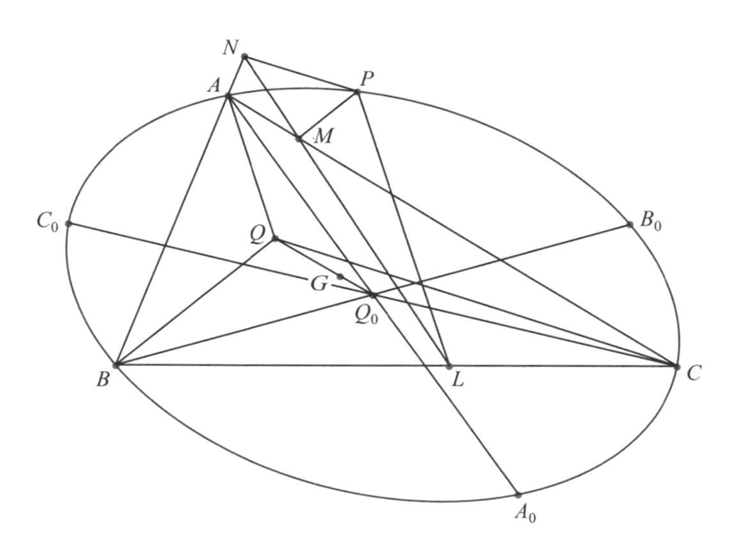

Fig. 10.2 The Simson conic of Q and the Simson line of P with respect to Q.

Exercise 10.2

10.2.1 If P lies on S_Q, the Simson conic of Q, then it is shown above that the midpoint X of PQ lies on the Simson line of P with respect to Q. Prove that the locus of X as P varies on S_Q is a conic Σ_Q. Find its equation in terms of the co-ordinates of Q, and prove the following.

(i) Σ_Q passes through L, M, and N, the midpoints of BC, CA, and AB, respectively.

(ii) Σ_Q passes through the midpoints of AQ, BQ, and CQ.

(iii) Σ_Q passes through the points D, E, and F on BC, CA, and AB, respectively, such that $BD/DC = w/v$, $CE/EA = u/w$, and $AF/FB = v/u$. Show also that its centre is the midpoint of QQ_0. Here (u, v, w) are the co-ordinates of Q.

10.3 The nine-point conic

This is the name of the conic Σ_Q, which is the subject of Exercise 10.2.1. The nine points are listed in that exercise. When Q is the orthocentre H, the nine-point conic becomes the nine-point circle.

From Exercise 10.2.1 the equation of Σ_Q is

$$vwx^2 + wuy^2 + uvz^2 - u(v+w)yz - v(w+u)zx - w(u+v)xy = 0\,. \quad (10.3.1)$$

When u, v, and w are rational we can find all points (x, y, z) on this conic with rational co-ordinates. Since eqn (10.3.1) is homogeneous, we can use unnormalised co-ordinates and suppose, therefore, that they are integral. In terms of two integer parameters p and q the solution is

$$\begin{aligned} x &= q(p+q)uv - q(p-q)wu\,, \\ y &= p(p+q)uv - p(q-p)vw\,, \\ z &= (p+q)\left[(p+q)uv + pvw + qwu\right]. \end{aligned} \quad (10.3.2)$$

For example, if we choose $(u, v, w) = (1/3, 1/2, 1/6)$ and $p = 3$ and $q = 2$ then we find the rational point $(x, y, z) = (28/185, 99/370, 43/74)$.

If Q is the centroid, then Σ_Q is the ellipse that touches the sides of ABC at its midpoints, with equation

$$x^2 + y^2 + z^2 - 2yz - 2zx - 2xy = 0\,. \quad (10.3.3)$$

The solution of this equation in terms of two parameters is $x = q^2$, $y = p^2$, and $z = (p+q)^2$. Here p and q may be taken to be any pair of integers, not both zero. This leads to the following rather curious theorem.

Theorem 10.3.1 *If $P(x, y, z)$ is a rational point on the ellipse touching the triangle ABC at its midpoints, then the ratios $[BCP] : [CAP] : [ABP]$ are rational squares.* □

Exercise 10.3

10.3.1 Let ABC be the right-angled triangle with $a = 3$, $b = 4$, and $c = 5$. Find the equations of S_Q and Σ_Q when Q is the incentre. In both cases find all points on these conics with rational co-ordinates.

11 Finite geometries

In a book devoted to the appearance of integers in geometry it seems fitting that a chapter should be devoted to a brief description of finite projective and finite affine geometries, since these produce configurations with finite numbers of points and lines.

11.1 Finite projective and affine geometries

In Section 4.2 it was shown how to embed $\mathbb{R}^2$ into the projective plane $\mathbb{P}^2$ by writing $x = X/Z$ and $y = Y/Z$, and making the equations of lines and curves homogeneous in X, Y, and Z. The points (x, y, z) and (kx, ky, kz) $(k \neq 0)$ are identified as the same point. In this way $Z = 0$ is regarded as the line at infinity, and the line $lX + mY + nZ = 0$ meets $Z = 0$ at a point P at infinity with co-ordinates $(m, -l, 0)$. Likewise, two lines which are parallel in $\mathbb{R}^2$, with equations $lx + my + n = 0$ and $lx + my + p = 0$ $(p \neq n)$ meet in $\mathbb{P}^2$ at P. This procedure served the purpose it was needed for in the sections on cubic curves and elliptic curves. The disadvantage of proceeding in this way is that points with $Z = 0$ are singled out as being something special, and one might rightly ask the question why X, Y, and Z are not on an equal footing.

In order to achieve the required generality it is best to start again, and produce an abstract projective plane, which subsequently may be associated with an affine plane by identifying *any* line in the projective plane with the line at infinity. The starting-point is a three-dimensional vector space V^3 defined over a commutative field F. The field F is sometimes taken to be the complex field, sometimes the real field, and for finite projective geometries it is a finite field, such as Z_p, the field of residues modulo p, where p is prime. For the moment we shall leave the field unspecified. The points of $\mathbb{P}^2$ are taken to be the rays of V^3, that is, the lines through the origin. So the points (x, y, z) and (kx, ky, kz), where k is a nonzero element of F, are the same point. The point $(0, 0, 0)$ is deleted. There is no origin in $\mathbb{P}^2$. We use capital letters for variables. Planes through the origin in V^3, with equations of the form $lX + mY + nZ = 0$, become the lines $\{l, m, n\}$ of $\mathbb{P}^2$, signifying the points that lie on this line. The point (x, y, z) lies on $lX + mY + nZ = 0$ if and only if $lx + my + nz = 0$. This condition may also be thought of as the condition that the line with co-ordinates $\{l, m, n\}$ passes through the point (x, y, z). The lines, strictly speaking, are rays of the dual space V^{3*}, and $\{l, m, n\}$ and $\{kl, km, kn\}$ represent the same line, where k is any nonzero element of F. The equation $xL + yM + zN = 0$ may be thought of as the equation of the point (x, y, z), signifying the lines that pass through this point. In this way there is a dual relationship in projective geometry between lines and points. For example, a quadrangle consists of four points and the six lines joining pairs of those points. A quadrilateral consists of four lines and the six points that are the intersections of pairs of those lines. Any theorem in projective geometry about a quadrangle will have its counterpart about a quadrilateral, and vice versa. In projective geometry there are no such things as parallel lines.

Theorem 11.1.1 *Every pair of distinct lines meet at a point, and a line joins every distinct pair of points.*

Proof The lines with equations $lX + mY + nZ = 0$ and $l'X + m'Y + n'Z = 0$ meet at the point with co-ordinates $(mn' - nm', nl' - ln', lm' - ml')$, and not all of these co-ordinates vanish simultaneously, the lines being distinct, so that $\{l', m', n'\} \neq \{kl, km, kn\}$ for any nonzero k. Likewise, the points with equations $xL+yM+zN = 0$ and $x'L+y'M+z'N = 0$ are joined by the line with co-ordinates $\{yz' - zy', zx' - xz', xy' - yx'\}$, and not all of these co-ordinates vanish simultaneously, the points being distinct, so that $(x', y', z') \neq (kx, ky, kz)$ for any nonzero k. □

We use the notation $\boldsymbol{A}$ to denote the co-ordinates (a_1, a_2, a_3) of the point A. The points A, B, and C are said to be linearly independent if the equation $a\boldsymbol{A}+b\boldsymbol{B}+c\boldsymbol{C} = \boldsymbol{0}$ implies that $a = b = c = 0$. Otherwise, when a, b, and c, not all zero, exist so that this equation holds, then A, B, and C are said to be *linearly dependent*. Analogous definitions hold for the dependency of lines. The points with co-ordinates $(1, 0, 0)$, $(0, 1, 0)$, and $(0, 0, 1)$ are clearly linearly independent and are often called the vertices of the triangle of reference. It has sides with equations $X = 0$, $Y = 0$, and $Z = 0$. The projective plane contains at least four points, no three of which are collinear, the vertices of the triangle of reference, and the point with co-ordinates $(1, 1, 1)$, since this point clearly does not lie on the sides of the triangle of reference. The maximum number of linearly independent points amongst any set of points in $\mathbb{P}^2$ is clearly three.

We now turn to the topic of finite projective geometries, that is, those geometries in which the number field has only a finite number of elements, and consequently there are just a finite number of points and lines. Firstly, we give in full detail the points and lines of $\mathbb{P}^2(2)$, where the field is Z_2, with arithmetic modulo 2. There are seven points with co-ordinates $A(1, 0, 0)$, $B(0, 1, 0)$, $C(0, 0, 1)$, $D(0, 1, 1)$, $E(1, 0, 1)$, $F(1, 1, 0)$, and $G(1, 1, 1)$. Dually, there are seven lines: $a\{1, 0, 0\}$ containing B, C, and D, $b\{0, 1, 0\}$ containing A, C, and E, $c\{0, 0, 1\}$ containing A, B, and F, $d\{0, 1, 1\}$ containing A, D, and G, $e\{1, 0, 1\}$ containing B, E, and G, $f\{1, 1, 0\}$ containing C, F, and G, and $g\{1, 1, 1\}$ containing D, E, and F. Likewise, b, c, and d are concurrent at A, since A has equation $L = 0$. Similarly, d, e, and f are concurrent at G, since G has equation $L + M + N = 0$. Further consideration shows that there are three lines through each point, as well as three points on each line. If we delete, for example, the line g with equation $x + y + z = 0$, then we are left with an affine geometry $A^2(2)$ consisting of the four points A, B, C, and G and the six lines a, b, c, d, e, and f. Their equations are $x = 0$, $y = 0$, $z = 0$, $y = z$, $z = x$, and $x = y$, respectively. The lines

$x = 0$ and $y = z$ are parallel, since in $\mathbb{P}^2(2)$ they meet at $D(0, 1, 1)$, which is on the deleted line. Similarly, $y = 0$ and $z = x$ are parallel, as are $z = 0$ and $x = y$.

An inability to draw convincing diagrams in these cases does not show that these geometries are in any sense invalid; it is simply that they contain generalisations of familiar concepts applied in (what nowadays are) unfamiliar contexts.

We next turn to the case of $\mathbb{P}^2(p)$, where the field is Z_p, the field of integers modulo p, where p is prime. Each point has three co-ordinates, which can range from 0 to $p-1$. However, not all of them can be zero. Also, each point is represented by $p-1$ multiples of a given triple. Hence the number of points is $(p^3-1)/(p-1) = p^2+p+1$. Dually, there are p^2+p+1 lines. In order to determine how many points there are on each line, we consider the line AB. It consists of all points of the form $a\boldsymbol{A} + b\boldsymbol{B}$. Apart from the point B, when $a = 0$, these may all be expressed in the form $\boldsymbol{A} + k\boldsymbol{B}$, where $k = a^{-1}b$, and distinct points arise when $k = 0, 1, \ldots, p-1$. It follows that there are $p+1$ points on each line and, dually, there are $p+1$ lines through each point.

Exercises 11.1

11.1.1 Prove that the equation of the line joining the points with co-ordinates $P_1(x_1, y_1, z_1)$ and $P_2(x_2, y_2, z_2)$ may be written as the vanishing of the determinant whose first row is (x, y, z) and whose second and third rows are the co-ordinates of P_1 and P_2.

11.1.2 Prove that in the finite affine geometry $A^2(p)$ there are p^2 points and p^2+p lines. Show further that these p^2+p lines form $p+1$ sets of p parallel lines.

11.1.3 List the points and lines of $\mathbb{P}^2(3)$. Delete the line $z = 0$, and hence list the points and lines of $A^2(3)$, indicating the four sets of three parallel lines.

11.1.4 A *difference-set* of integers mod n^2+n+1 is a set of $n+1$ distinct integers such that the $n(n+1)$ signed differences between pairs of the set yield every integer mod n^2+n+1 except zero.

(a) Find difference-sets mod 7, mod 13, and mod 21.

(b) Show that, if $x_1, x_2, \ldots, x_{n+1}$ is a difference-set mod n^2+n+1, then one can construct a projective plane $\mathbb{P}^2(n)$ by labelling the points P_j, $j = 0, \ldots, n^2+n$, and the lines p_k, $k = 0, \ldots, n^2+n$, with the rule that P_j and p_k are incident if and only if $j + k = x_t \pmod{n^2+n+1}$, for some $t = 1, \ldots, n+1$. Hence investigate $\mathbb{P}^2(4)$.

Appendix A Areal co-ordinates

Several sections on the triangle, in particular, Sections 6.3 and 6.4, the whole of Chapter 8, and Sections 10.2 and 10.3, rely on a knowledge of areal (or barycentric) co-ordinates. As far as I am aware, there is no recent textbook that gives a comprehensive account of what they are and how useful they are in dealing with a variety of geometrical problems involving a triangle and the circles connected with a triangle. Some books give a short treatment of them, but do not go into much detail. For example, Coxeter (1989) and Silvester (2001) define areal co-ordinates and use them in dealing with the theorems of Ceva and Menelaus. However, as far as I am aware, there is no indication in any modern text of how they can be used to calculate distances between key points in a triangle or how they can be used to solve complicated problems involving the interaction between a triangle and certain lines, circles, and conic sections.

I shall therefore give a fairly full account of areal co-ordinates, both to give the reader the necessary background for the above-mentioned sections and to fill what appears to be a gap in the literature. Some readers may be well acquainted with areal co-ordinates or with trilinear co-ordinates, which are their closest relation. Their application to problems concerning incidence, such as whether three points are collinear or not, is similar to the use of homogeneous co-ordinates in projective geometry. However, it may not be so widely known that a metric exists, which renders them useful in many problems in Euclidean geometry.

A.1 Preliminaries

We assume a basic knowledge of vectors, including the scalar and vector product. If O is the origin and P is any other point then we usually denote the position vector $\boldsymbol{OP}$ by $\boldsymbol{p}$. The exceptions are in dealing with the vertices A, B, and C of a triangle ABC when we write $\boldsymbol{OA} = \boldsymbol{x}$, $\boldsymbol{OB} = \boldsymbol{y}$, and $\boldsymbol{OC} = \boldsymbol{z}$. The reason for this is that we reserve the symbols a, b, and c for the side lengths BC, CA, and AB, respectively, and we wish to avoid any confusion between the shortened form of the symbol $|\boldsymbol{OA}|$ and the side BC. Note that O is not usually the circumcentre of the triangle ABC, but is an arbitrary origin, unless we happen to say otherwise and give its location.

A.2 The co-ordinates of a line

We start by revising some ideas about points on a line. If P and Q are two points on a line L, then, using vectors, the equation of the line PQ is given by

$$\boldsymbol{r} = \boldsymbol{p} + t(\boldsymbol{q} - \boldsymbol{p}) = (1-t)\boldsymbol{p} + t\boldsymbol{q}\,. \tag{A.2.1}$$

Here t is a parameter such that $-\infty < t < \infty$, its value depending on where R lies on PQ. For example, when $t = 0$ the point R coincides with P, and when $t = 1$ the point R coincides with Q. Points with $0 < t < 1$ lie on the line segment PQ; points with $t < 0$ lie on the extension of QP beyond P; and points with $t > 1$ lie on the extension of PQ beyond Q. If we put $t = m/(l+m)$, where $l + m \neq 0$, then the equation of L becomes

$$(l+m)\boldsymbol{r} = l\boldsymbol{p} + m\boldsymbol{q}\,. \tag{A.2.2}$$

For example, when $l = 1$ and $m = 2$, then $\boldsymbol{r} = (1/3)(\boldsymbol{p} + 2\boldsymbol{q})$ and R lies on the line segment PQ and is such that $\boldsymbol{PR} = 2\boldsymbol{RQ}$. When $l = 1$ and $m = 0$ the point R lies at P, and when $l = 0$ and $m = 1$ the point R lies at Q.

From now on we shall always insist that $l + m = 1$; the equation of PQ then becomes

$$\boldsymbol{r} = l\boldsymbol{p} + m\boldsymbol{q}\,, \tag{A.2.3}$$

and we may say that every point R on the line has a pair of co-ordinates (l, m) subject to the condition $l + m = 1$, where $(1, 0)$ represents P, $(0, 1)$ represents Q, and a point such as $(1/3, 2/3)$ represents a point R such that $PR/RQ = 2$. More generally, if $l > 0$ and $m > 0$, then R lies on the line segment between P and Q and is such that $PR/RQ = m/l$. Points on the extension of PQ beyond Q have co-ordinates (l, m) such that $m > 1$ and $l < 0$. If we now give lengths a sign, so that those lengths in the direction from P to Q are positive, and those in the direction from Q to P are negative, then the equation $PR/RQ = m/l$ still holds, since PR is positive and RQ is negative. The same sort of argument holds for points on the extension of QP beyond P. For example, when R has co-ordinates $(2, -1)$ we have $PR/RQ = -\frac{1}{2}$, with PR negative and RQ positive. The point R with co-ordinates $(1/3, 2/3)$ is said to divide PQ *internally* in the ratio $2 : 1$, whereas the point R with co-ordinates $(2, -1)$ is said to divide PQ *externally* in the ratio $1 : 2$.

Theorem A.2.1 *The points P, Q, and R are collinear if and only if there exist constants k, l, and m, not all zero, such that $k\boldsymbol{r} + l\boldsymbol{p} + m\boldsymbol{q} = \boldsymbol{0}$ and $k + l + m = 0$.*

Proof If R lies on PQ then, by eqn (A.2.3), we can choose $k = -1$ and l and m such that $l + m = 1$.

Conversely, if $k\boldsymbol{r} + l\boldsymbol{p} + m\boldsymbol{q} = \boldsymbol{0}$ and $k + l + m = 0$, then, since not all of k, l, and m are zero, one of them is certainly not zero, and we may suppose, without loss of generality, that $k \neq 0$. By scaling, we may ensure that $k = -1$. Then $\boldsymbol{r} = l\boldsymbol{p} + m\boldsymbol{q}$ and, by eqn (A.2.3), R lies on PQ. □

Theorem A.2.2 *The co-ordinate system (l, m) for the line L, given by eqn (A.2.3), is independent of the position of the origin O.*

Proof Suppose that a second origin O' is taken with $\boldsymbol{OO'} = \boldsymbol{d}$. Then, using an obvious notation, with primed quantities denoting vector positions with respect to O', we have $\boldsymbol{p} = \boldsymbol{d} + \boldsymbol{p}'$, $\boldsymbol{q} = \boldsymbol{d} + \boldsymbol{q}'$, and $\boldsymbol{r} = \boldsymbol{d} + \boldsymbol{r}'$, and eqn (A.2.3) transforms into

$$\boldsymbol{d} + \boldsymbol{r}' = l(\boldsymbol{d} + \boldsymbol{p}') + m(\boldsymbol{d} + \boldsymbol{q}') . \tag{A.2.4}$$

However, $l + m = 1$ and so eqn (A.2.4) becomes

$$\boldsymbol{r}' = l\boldsymbol{p}' + m\boldsymbol{q}' . \tag{A.2.5}$$

The co-ordinates (l, m) are therefore unlike rectangular Cartesian co-ordinates, which alter when the origin is translated. □

So far, we have an algebraic representation of points on a line in which we have *signed ratios*. To summarise, if R has line co-ordinates (l, m), where $l + m = 1$, and lies on the line PQ, where P has co-ordinates $(1, 0)$ and Q has co-ordinates $(0, 1)$, then $PR/RQ = m/l$, where line segments are signed, as described above.

We now show how to introduce a metric on the line. It may seem at first sight to be contrived and more elaborate than necessary, but it turns out to be the correct form for generalisation into higher dimensions. Let the length of PQ be c. We now suppose that R and S are two points on the line PQ with line co-ordinates $R(l, m)$ and $S(r, s)$, where $l + m = 1$ and $r + s = 1$. Then $\boldsymbol{RS} = (r - l, s - m) = (u, v)$, where $u = r - l$ and $v = s - m$, and hence $u + v = 0$. Note that, for any displacement such as $\boldsymbol{RS}$ along the line PQ, the sum of the co-ordinates is zero. Now $\boldsymbol{PQ} = (-1, 1)$ and we know, by definition, that the length of PQ is c. This is at least consistent with the possibility that the required metric is given by the equation

$$\boldsymbol{RS}^2 = -c^2 uv . \tag{A.2.6}$$

In fact, this is easily confirmed, since $\boldsymbol{RS} = (u, v) = (-v, v) = v(-1, 1) = v\boldsymbol{PQ}$, and hence $\boldsymbol{RS}^2 = v^2\boldsymbol{PQ}^2 = -c^2uv$. This may at first appear contrived; but it is the form that generalises when we come to deal with areal co-ordinates and distances in the plane of a triangle.

A.3 The vector treatment of a triangle

Let ABC be a triangle in the plane Π. Let O be an arbitrary origin and suppose that $\boldsymbol{OA} = \boldsymbol{x}$, $\boldsymbol{OB} = \boldsymbol{y}$, and $\boldsymbol{OC} = \boldsymbol{z}$. We quote, without proof, that the signed area of the triangle ABC is given by

$$[ABC]\boldsymbol{k} = \frac{1}{2}(\boldsymbol{y} \times \boldsymbol{z} + \boldsymbol{z} \times \boldsymbol{x} + \boldsymbol{x} \times \boldsymbol{y}) , \tag{A.3.1}$$

where $\boldsymbol{k}$ is a *fixed* unit normal perpendicular to Π. The choice of the direction of $\boldsymbol{k}$ dictates the sign of $[ABC]$. This now creates a sign convention for areas in the plane

Π. Thus, if $[ABC]$ is positive and P lies on the same side of BC as A then $[PBC]$ is positive, but if P lies on the other side of BC from A then $[PBC]$ is negative.

Suppose now that O lies external to Π. It is well known that $\boldsymbol{x}$, $\boldsymbol{y}$, and $\boldsymbol{z}$ are linearly independent and that $\boldsymbol{x} \cdot (\boldsymbol{y} \times \boldsymbol{z}) \neq 0$, since it is the signed volume V of the parallelepiped based on OA, OB, and OC as base vectors. (In fact, the vanishing of V, when none of $\boldsymbol{x}$, $\boldsymbol{y}$, or $\boldsymbol{z}$ is the zero vector, is a necessary and sufficient condition for A to lie in the plane OBC, which would mean that ABC is a degenerate triangle.) Note that, together with eqn (A.3.1), this implies that

$$[ABC]\boldsymbol{k} \cdot \boldsymbol{x} = [ABC]\boldsymbol{k} \cdot \boldsymbol{y} = [ABC]\boldsymbol{k} \cdot \boldsymbol{z} = \frac{1}{2}V\,. \tag{A.3.2}$$

Theorem A.3.1 *If O does not lie in the plane Π of a triangle ABC, then, for any point $P \in \Pi$, there exist unique constants l, m, and n such that the position vector of P is given by*

$$\boldsymbol{p} = l\boldsymbol{x} + m\boldsymbol{y} + n\boldsymbol{z} \quad \textit{and} \quad l + m + n = 1\,. \tag{A.3.3}$$

Proof Using oblique axes AB and AC in Π, we have $\boldsymbol{OP} = \boldsymbol{OA} + \alpha\boldsymbol{AB} + \beta\boldsymbol{AC}$, that is,

$$\boldsymbol{p} = \boldsymbol{x} + \alpha(\boldsymbol{y} - \boldsymbol{x}) + \beta(\boldsymbol{z} - \boldsymbol{x})\,. \tag{A.3.4}$$

Now take $l = 1 - \alpha - \beta$, $m = \alpha$, $n = \beta$, and $l + m + n = 1$, and we obtain eqn (A.3.3). Uniqueness follows from the linear independence of $\boldsymbol{x}$, $\boldsymbol{y}$, and $\boldsymbol{z}$. For, if $\boldsymbol{p} = l'\boldsymbol{x} + m'\boldsymbol{y} + n'\boldsymbol{z}$, with $l' + m' + n' = 1$ and $(l', m', n') \neq (l, m, n)$, then $u\boldsymbol{x} + v\boldsymbol{y} + w\boldsymbol{z} = \boldsymbol{0}$, where $u = l - l'$, $v = m - m'$, and $w = n - n'$, and not all of u, v, and w are zero. This would contradict the linear independence of $\boldsymbol{x}$, $\boldsymbol{y}$, and $\boldsymbol{z}$. Hence the representation (A.3.3) is unique. □

However, we do not wish to restrict the position of O, and the nice thing is that we need not do so.

Theorem A.3.2 *Even if O lies in the plane Π of the triangle ABC, then, for any point $P \in \Pi$, there exist unique constants l, m, and n such that $l + m + n = 1$ and the position vector of P is given by eqn (A.3.3).*

Proof In this case, since $\boldsymbol{x}$, $\boldsymbol{y}$, and $\boldsymbol{z}$ are not now independent, there exist constants r, s, and t, not all zero, such that $r\boldsymbol{x} + s\boldsymbol{y} + t\boldsymbol{z} = \boldsymbol{0}$, and, since A, B, and C are not collinear, by Theorem A.2.1, we have $r+s+t = d \neq 0$. Now let $\boldsymbol{p} = l_0\boldsymbol{x} + m_0\boldsymbol{y} + n_0\boldsymbol{z}$, where $l_0 + m_0 + n_0 = c$. Such expansions exist and, with c allowed to vary, they are not unique. Now define k such that $c - kd = 1$. Then

$$\boldsymbol{p} = l_0\boldsymbol{x} + m_0\boldsymbol{y} + n_0\boldsymbol{z} - k(r\boldsymbol{x} + s\boldsymbol{y} + t\boldsymbol{z}) = l\boldsymbol{x} + m\boldsymbol{y} + n\boldsymbol{z}\,, \tag{A.3.5}$$

where $l + m + n = (l_0 + m_0 + n_0) - k(r + s + t) = c - kd = 1$. Suppose now that there is a second representation $\boldsymbol{p} = l'\boldsymbol{x} + m'\boldsymbol{y} + n'\boldsymbol{z}$, where $l' + m' + n' = 1$ and $(l', m', n') \neq (l, m, n)$. Then, writing $u = l - l'$, $v = m - m'$, and $w = n - n'$, we have $u\boldsymbol{x} + v\boldsymbol{y} + w\boldsymbol{z} = \boldsymbol{0}$ and $u + v + w = 0$. However, by Theorem A.2.1, this means that A, B, and C are collinear. Since this is not so, the representation (A.3.5) is unique. □

Theorem A.3.3 *The co-ordinate system (l, m, n) for the plane Π, defined by eqns (A.3.3) and (A.3.5), is independent of the position of the origin O.* □

The proof is left to the reader, as it is identical, apart from the dimension, to the proof of Theorem A.2.2.

A.4 Why the co-ordinates (l, m, n) are called areal co-ordinates

The co-ordinates (l, m, n) of points P in the plane of Π, given by $\boldsymbol{p} = l\boldsymbol{x} + m\boldsymbol{y} + n\boldsymbol{z}$, with $l + m + n = 1$, are called areal co-ordinates, for reasons which we now explain. First note, however, that $(1, 0, 0)$ represents A, with $\boldsymbol{p} = \boldsymbol{x}$, $(0, 1, 0)$ represents B, with $\boldsymbol{p} = \boldsymbol{y}$, and $(0, 0, 1)$ represents C, with $\boldsymbol{p} = \boldsymbol{z}$. For this reason ABC is often called the *triangle of reference*.

Next, recall eqn (A.3.1) which states that $[ABC]\boldsymbol{k} = \frac{1}{2}(\boldsymbol{y} \times \boldsymbol{z} + \boldsymbol{z} \times \boldsymbol{x} + \boldsymbol{x} \times \boldsymbol{y})$. Let us now calculate $[PBC]\boldsymbol{k}$, where $\boldsymbol{p} = l\boldsymbol{x} + m\boldsymbol{y} + n\boldsymbol{z}$. We have

$$\begin{aligned}[PBC]\boldsymbol{k} &= \frac{1}{2}(\boldsymbol{BC} \times \boldsymbol{BP}) \\ &= \frac{1}{2}(\boldsymbol{z} - \boldsymbol{y}) \times [l\boldsymbol{x} + (m-1)\boldsymbol{y} + n\boldsymbol{z}] \\ &= \frac{1}{2}l(\boldsymbol{y} \times \boldsymbol{z} + \boldsymbol{z} \times \boldsymbol{x} + \boldsymbol{x} \times \boldsymbol{y}) \\ &= l[ABC]\boldsymbol{k}\,. \end{aligned} \tag{A.4.1}$$

Hence

$$l = \frac{[PBC]}{[ABC]}\,. \tag{A.4.2}$$

Similarly,

$$m = \frac{[APC]}{[ABC]} \tag{A.4.3}$$

and

$$n = \frac{[ABP]}{[ABC]}\,. \tag{A.4.4}$$

The sign convention, explained in Section A.3, means that wherever P is situated we have $[PBC] + [APC] + [ABP] = [ABC]$, which is the geometrical expression, in terms of areas, of the condition $l + m + n = 1$. This is why the co-ordinates are called areal co-ordinates, since, by virtue of the uniqueness, there are no other sets of co-ordinates having the property that $l + m + n = 1$.

When $l = 0$ the point P lies on BC, and, since $m + n = 1$, (m, n) are the line co-ordinates for points on BC. Similarly, the co-ordinates (n, l) and (l, m) are the line co-ordinates for points on CA and AB, respectively. See Fig. A.1 for a diagram showing the regions where l, m, and n are positive or negative. The notation in Fig. A.1 means that $+--$, for example, indicates a region in which l is positive and m and n are negative. There are seven regions formed by the sides of a triangle and their extensions, and all selections are represented except $---$.

Note that, in working out an area in Π, the area appears equally as the coefficient of $\boldsymbol{y} \times \boldsymbol{z}$, $\boldsymbol{z} \times \boldsymbol{x}$, and $\boldsymbol{x} \times \boldsymbol{y}$, which means that, in practice, only one of the coefficients needs to be calculated.

After such a thorough explanation of what areal co-ordinates are, you might imagine that we are ready to do some useful geometry. Unfortunately, we are not quite there yet, since we do not even know how to write down the equation of a line, nor even that it is linear in the co-ordinates of a variable point on the line. We use as the co-ordinates of a variable point the triple (x, y, z), and use other triples such as (l, m, n) for the co-ordinates of fixed points. Note that in this context x, y, and z are nothing to do with the lengths of $\boldsymbol{x}$, $\boldsymbol{y}$, and $\boldsymbol{z}$. An account is therefore necessary of how to obtain the equation of a line, and, as the account provides a formula for the area of an arbitrary triangle in the plane of ABC, there is a bonus in proceeding in this way.

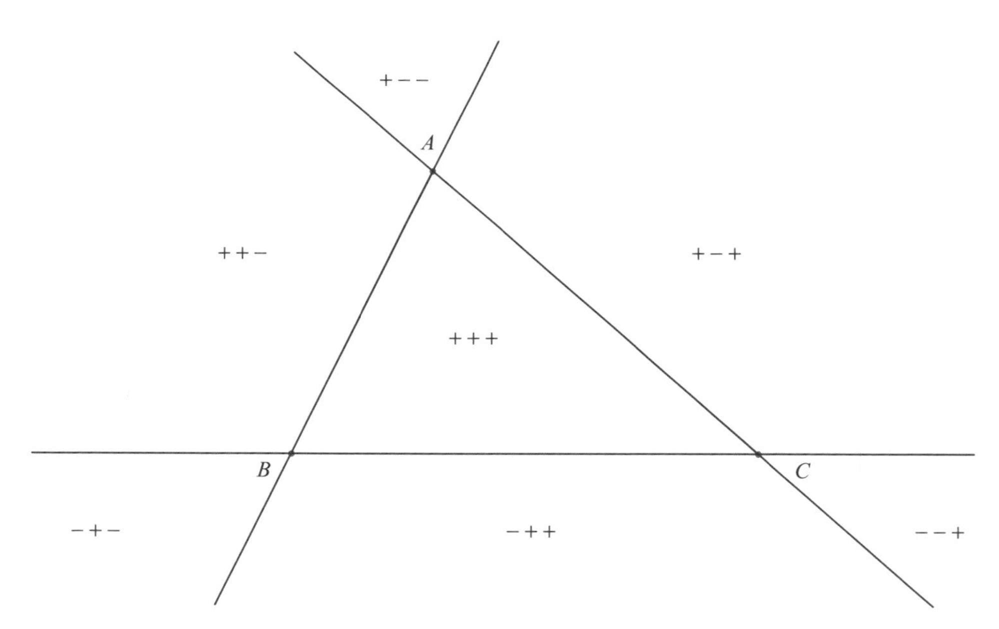

Fig. A.1 The signs of l, m, and n in the various regions.

A.5 The area of a triangle PQR and the equation of the line PQ

Suppose that P, Q, and R have areal co-ordinates (d, e, f), (g, h, k), and (l, m, n), respectively. As remarked above, in working out an area in Π, since it appears equally as the coefficient of $\boldsymbol{y} \times \boldsymbol{z}$, $\boldsymbol{z} \times \boldsymbol{x}$, and $\boldsymbol{x} \times \boldsymbol{y}$, only one of these coefficients needs to be calculated. It follows that

$$
\begin{aligned}
\frac{[PQR]}{[ABC]} &= \text{the first co-ordinate in the vector product} \\
&\qquad\qquad (l-g, m-h, n-k) \times (d-g, e-h, f-k) \\
&= (m-h)(f-k) - (n-k)(e-h) \\
&= mf - mk - hf - ne + nh + ke \\
&= (hn - mk) + (mf - en) + (ek - hf) \\
&= (1-e-f)(hn-mk) \\
&\qquad + (1-h-k)(mf-en) + (1-m-n)(ek-hf) \\
&= d(hn-mk) + g(mf-en) + l(ek-hf) \\
&= \det \begin{vmatrix} d & g & l \\ e & h & m \\ f & k & n \end{vmatrix}. \qquad \text{(A.5.1)}
\end{aligned}
$$

In other words, the ratio of the signed areas of the triangles PQR and ABC is the determinant of the matrix whose columns are, respectively, the co-ordinates of the points P, Q, and R.

The line PQ is the locus of points R such that $[PQR]/[ABC] = 0$, and therefore has equation

$$x(ek - fh) + y(fg - dk) + z(dh - eg) = 0. \qquad \text{(A.5.2)}$$

This is, of course, exactly what you would expect if you are familiar with the equation of a line using homogeneous co-ordinates in projective geometry. It is linear, homogeneous of degree 1 in x, y, and z, and clearly contains the points P and Q.

There are some immediate consequences of eqn (A.5.2). Firstly, note that $ek - fh$, $fg - dk$, and $dh - eg$ are equal to $[APQ]/[ABC]$, $[BPQ]/[ABC]$, and $[CPQ]/[ABC]$, respectively, and, since the triangles APQ, BPQ, and CPQ have a common side PQ, we have the following theorem.

Theorem A.5.1 *Given a line with equation $px+qy+rz = 0$, the signed perpendicular distances from A, B, and C onto the line are proportional to p, q, and r, respectively.* □

If an arbitrary point R has co-ordinates (l, m, n) and p, q, and r are the actual perpendicular distances from A, B, and C, respectively, onto the line, then it follows immediately from eqn (A.5.1) that $|pl + qm + rn|$ is the perpendicular distance from R onto the line PQ.

Theorem A.5.2 *The condition for the two (non-identical) lines with equations* $p_1x + q_1y + r_1z = 0$ *and* $p_2x + q_2y + r_2z = 0$ *to be parallel is*

$$q_1r_2 - q_2r_1 + r_1p_2 - r_2p_1 + p_1q_2 - p_2q_1 = 0\,. \tag{A.5.3}$$

Proof The two lines are parallel if they do not meet at a real point. Now real points correspond to those triples (x, y, z) that can be normalised so that $x + y + z = 1$. Hence the lines are parallel if their equations, together with the equation $x+y+z = 0$, have a solution in which x, y, and z are not all zero. (Such points may be thought of as lying on the line at infinity.) This is so if and only if

$$\det \begin{vmatrix} 1 & 1 & 1 \\ p_1 & q_1 & r_1 \\ p_2 & q_2 & r_2 \end{vmatrix} = 0\,. \tag{A.5.4}$$

Equation (A.5.3) is the expansion of the condition (A.5.4). □

We are now in a position to do some geometry, provided that actual distances are not involved. We postpone the derivation of the areal metric until after examples illustrating the theory so far. Without a metric we are limited to results involving points and lines, and incidence properties, such as collinearity and concurrence. Metric properties, including conditions for lines to be at right angles, are given in Section A.8.

A.6 The areal co-ordinates of key points in the triangle

The centroid G

We have already seen that A, B, and C have co-ordinates $A(1,0,0)$, $B(0,1,0)$, and $C(0,0,1)$. The equations of the lines BC, CA, and AB are therefore $x = 0$, $y = 0$, and $z = 0$, respectively. Therefore the midpoints L, M, and N of the sides BC, CA, and AB, respectively, have co-ordinates $L(0, \frac{1}{2}, \frac{1}{2})$, $M(\frac{1}{2}, 0, \frac{1}{2})$, and $N(\frac{1}{2}, \frac{1}{2}, 0)$. It follows that the equations of the medians AL, BM, and CN are $y = z$, $z = x$, and $x = y$, respectively. These are concurrent at the centroid G, where $x = y = z = 1/3$. Its co-ordinates are $G(1/3, 1/3, 1/3)$. This implies the well-known fact that $[GBC] = [AGC] = [ABG] = (1/3)[ABC]$.

The incentre I

The incentre I has the property that the perpendicular distances from I to all three sides are equal to the inradius. It follows that $[IBC] : [AIC] : [ABI] = a : b : c$. Hence the co-ordinates of the incentre are $I(a, b, c)/(a + b + c)$. The equations of the lines AI, BI, and CI are therefore $cy = bz$, $az = cx$, and $bx = ay$, respectively.

Theorem A.6.1 *The internal bisector of* $\angle BAC$ *meets* BC *at a point* U *such that* $BU/UC = AB/CA$.

Proof AI has equation $cy = bz$ and meets BC with equation $x = 0$ at the point U with co-ordinates $(0, b, c)/(b + c)$. From our knowledge of the properties of line co-ordinates, this means that $BU/UC = c/b = AB/CA$. □

The circumcentre O

Let R be the circumradius; then $BC = a = 2R \sin A$ and $OL = R \cos A$. It follows that $[OBC] = R^2 \sin A \cos A = \frac{1}{2}R^2 \sin 2A$. Hence

$$[OBC] : [AOC] : [ABO] = \sin 2A : \sin 2B : \sin 2C\,. \qquad \text{(A.6.1)}$$

Since $\sin 2A + \sin 2B + \sin 2C = 4 \sin A \sin B \sin C$, the normalised co-ordinates of the circumcentre O are $O(\sin 2A, \sin 2B, \sin 2C)/4 \sin A \sin B \sin C$. Thus AO meets BC at a point P such that $BP/PC = \sin 2C/ \sin 2B$.

The orthocentre H

Let R be the circumradius, and let AH, BH, and CH meet the sides BC, CA, and AB at D, E, and F, respectively. It is easily calculated that $HD = 2R \cos B \cos C$ and $BC = 2R \sin A$. Hence $[HBC] = 2R^2 \sin A \cos B \cos C$. Since $[ABC] = 2R^2 \sin A \sin B \sin C$, it follows that the co-ordinates of H are

$$H(\cot B \cot C, \cot C \cot A, \cot A \cot B)\,.$$

This is a geometrical proof of the trigonometrical identity that when $A+B+C = 180°$,

$$\cot B \cot C + \cot C \cot A + \cot A \cot B = 1\,. \qquad \text{(A.6.2)}$$

It is often preferable to leave these co-ordinates in their unnormalised form $H(\tan A, \tan B, \tan C)$.

The more complicated forms for the co-ordinates of O and H make the use of areal co-ordinates for problems involving these points less attractive.

It is left as an exercise for the reader to prove that the excentres I_1, I_2, and I_3 of the excircles opposite A, B, and C, respectively, have co-ordinates

$$I_1 \frac{(-a, b, c)}{b + c - a}\,, \qquad I_2 \frac{(a, -b, c)}{c + a - b}\,, \qquad I_3 \frac{(a, b, -c)}{a + b - c}\,.$$

A.7 Some examples

Example A.7.1 (Menelaus' theorem) Suppose that the transversal LMN, with L, M, and N on BC, CA, and AB, respectively, has equation $px + qy + rx = 0$, where x, y, and z are current co-ordinates on the line and p, q, and r are constants. The line meets BC, whose equation is $x = 0$, at L with co-ordinates $(0, -r, q)/(q - r)$. (If $q = r$ then the transversal is parallel to BC and L does not exist.) Since these are now the line co-ordinates of L on BC, we have $BL/LC = -q/r$. Similarly, we have $CM/MA = -r/p$ and $AN/NB = -p/q$. It follows that

$$\frac{BL}{LC}\frac{CM}{MA}\frac{AN}{NB} = -1\,, \tag{A.7.1}$$

which is Menelaus' theorem for transversals of a triangle.

For the converse, if eqn (A.7.1) holds, let LM meet AB at N'. Then, by Menelaus' theorem, we have $(BL/LC)(CM/MA)(AN'/N'B) = -1$, so that $AN/NB = AN'/N'B$ and hence N and N' coincide.

Example A.7.2 (Ceva's theorem) Suppose that $P(l, m, n)$ is a point in the plane of the triangle ABC. Then the equation of the Cevian AP is $ny = mz$, and this meets BC at a point L with unnormalised co-ordinates $(0, m, n)$, and we have $BL/LC = n/m$. If the other two Cevians BP and CP meet CA and AB, respectively, at M and N, then, by a similar argument, $CM/MA = l/n$ and $AN/NB = m/l$. Hence

$$\frac{BL}{LC}\frac{CM}{MA}\frac{AN}{NB} = +1\,, \tag{A.7.2}$$

which is Ceva's theorem for the intercepts on the sides of a triangle by the Cevians. Note that P does not have to be internal to the triangle ABC, as the signed lengths in eqn (A.7.2) take care of any negative quantities amongst l, m, and n.

For the converse, it is possible for eqn (A.7.2) to hold with L, M, and N on BC, CA, and AB, respectively, but for AL, BM, and CN to be parallel. Excluding this case, we may suppose, without loss of generality, that AL and BM meet at P. Now let CP meet AB at N'. Then, by Ceva's theorem, we have

$$\frac{BL}{LC}\frac{CM}{MA}\frac{AN'}{N'B} = +1\,,$$

and so $AN'/N'B = AN/NB$ and N' coincides with N; thus AL, BM, and CN are concurrent at P.

Example A.7.3 (Gergonne's point) Suppose that the incircle touches BC, CA, and AB at X, Y, and Z, respectively. Then we know from elementary considerations that $BX = s - b$ and $XC = s - c$, where $s = \frac{1}{2}(a + b + c)$. It follows that the unnormalised co-ordinates of X are $(0, s - c, s - b)$. The equation of AX is therefore $(s - c)z = (s - b)y$. It is evident that the point P with unnormalised co-ordinates $P(1/(s - a), 1/(s - b), 1/(s - c))$ lies on this line. By symmetry, it is clear that P also lies on BY and CZ. It follows that AX, BY, and CZ are concurrent at P. This point is called Gergonne's point.

Note that, in dealing with the equations of lines, because those equations are homogeneous of degree 1 in x, y, and z, it is possible to use the unnormalised co-ordinates of points. This shortens and simplifies an argument such as the one above.

A.8 The areal metric

Although I was somewhat surprised when I discovered that a metric exists, perhaps I should not have been. There is, after all, something about an area that is suggestive of measurement. Areas themselves appear to be measured relative to the area of the triangle ABC; so surely distances ought somehow to be measured relative to the side lengths a, b, and c of the triangle? This, in fact, turns out to be the case. However, the derivation is not easy, and the argument has to be broken down into three cases, namely, when triangle ABC is acute, when it is right-angled, and when it is obtuse. Of course, the same formula emerges in all three cases.

- *Case 1, ABC is an acute-angled triangle*
  Since areal co-ordinates are independent of the choice of origin, we may choose any suitable origin. For an acute-angled triangle there exists a suitable origin O above the plane of the triangle such that $\boldsymbol{OA}$, $\boldsymbol{OB}$, and $\boldsymbol{OC}$ form a right triad of vectors. If we take OA, OB, and OC to be in the x-, y-, and z-directions, respectively, then the plane Π in which the triangle ABC lies has the equation in Cartesian co-ordinates given by $x+y+z=1$. Since areal co-ordinates are unique, they therefore coincide with Cartesian co-ordinates which are set up in this way.
  Why, you may ask, do we set up areal co-ordinates with all the elaborate machinery involving ratios if they are only Cartesian co-ordinates in disguise? The answer to this is that such a disguise is only possible for acute-angled triangles. No origin with $\boldsymbol{OA}$, $\boldsymbol{OB}$, and $\boldsymbol{OC}$ being a right triad of vectors exists for a right-angled or an obtuse-angled triangle. For these triangles it is true that the co-ordinates satisfy $x+y+z=1$, but, although it is possible to treat the right-angled triangle as a limiting case of an acute-angled triangle, it is not possible, given an obtuse-angled triangle, to choose any rectangular Cartesian co-ordinate system in which x, y, and z are the co-ordinates.
  Firstly, we must prove the existence of the point O with the required property. In an acute-angled triangle, by the cosine rule, all of the quantities $b^2+c^2-a^2$, $c^2+a^2-b^2$, and $a^2+b^2-c^2$ are positive. We may therefore set these quantities equal to $2d^2$, $2e^2$, and $2f^2$, respectively. Then

$$\begin{aligned} e^2+f^2 &= a^2\,, \\ f^2+d^2 &= b^2\,, \\ d^2+e^2 &= c^2\,, \end{aligned} \tag{A.8.1}$$

  and, by the converse of Pythagoras' theorem, a point O exists in which OA, OB, and OC are mutually at right angles and $OA=d$, $OB=e$, and $OC=f$. This construction only works for an acute-angled triangle. Since $\boldsymbol{OA}=\boldsymbol{x}$, $\boldsymbol{OB}=\boldsymbol{y}$, and $\boldsymbol{OZ}=\boldsymbol{z}$, we have

$$|\boldsymbol{x}|=d\,,\quad |\boldsymbol{y}|=e\,,\quad |\boldsymbol{z}|=f\,,\quad \boldsymbol{y}\cdot\boldsymbol{z}=\boldsymbol{z}\cdot\boldsymbol{x}=\boldsymbol{x}\cdot\boldsymbol{y}=0\,. \tag{A.8.2}$$

  Suppose now that we have two points P and Q in the plane Π with areal co-ordinates $P(l,m,n)$ and $Q(r,s,t)$. Then $\boldsymbol{PQ}=(r-l,s-m,t-n)=(u,v,w)$, where

$u = r - l, v = s - m, w = t - n$, and $u + v + w = 0$. Note that, for any displacement vector, the sum of the co-ordinates is zero. Now, since $\boldsymbol{PQ} = u\boldsymbol{x} + v\boldsymbol{y} + w\boldsymbol{z}$, we may use eqns (A.8.2) to give

$$\begin{aligned} PQ^2 &= u^2 d^2 + v^2 e^2 + w^2 f^2 \\ &= \frac{1}{2}(b^2 + c^2 - a^2)u^2 + \frac{1}{2}(c^2 + a^2 - b^2)v^2 + \frac{1}{2}(a^2 + b^2 - c^2)w^2 \\ &= \frac{1}{2}\left[a^2(v^2 + w^2 - u^2) + b^2(w^2 + u^2 - v^2) + c^2(u^2 + v^2 - w^2)\right] \\ &= -a^2 vw - b^2 wu - c^2 uv\,, \end{aligned} \tag{A.8.3}$$

since $u + v + w = 0$. The formula (A.8.3) is the long-awaited metric for a straight-line displacement $\boldsymbol{PQ} = (u, v, w)$, with $u + v + w = 0$.
A number of points need to be made. Firstly, eqn (A.8.3) generalises eqn (A.2.6) for the metric in line co-ordinates. Secondly, although it looks negative, it is in fact a positive-definite quadratic form (because of the condition $u + v + w = 0$). Thirdly, if $\mathbf{d}\boldsymbol{s} = (\mathrm{d}u, \mathrm{d}v, \mathrm{d}w)$ is an infinitesimal displacement then

$$\mathrm{d}s^2 = -a^2\,\mathrm{d}v\,\mathrm{d}w - b^2\,\mathrm{d}w\,\mathrm{d}u - c^2\,\mathrm{d}u\,\mathrm{d}v\,. \tag{A.8.4}$$

- *Case 2, ABC is a right-angled triangle*
  Without loss of generality, suppose that the right angle is at A. The suitable choice of origin is now at A itself. Although $\boldsymbol{OA} = \mathbf{0}$, $\boldsymbol{OB} = \boldsymbol{AB} = \boldsymbol{y}$, and $\boldsymbol{OC} = \boldsymbol{AC} = \boldsymbol{z}$, we may take this case to be a limiting form of Case 1. We now have $|\boldsymbol{x}|^2 = d^2 = \frac{1}{2}(b^2 + c^2 - a^2) = 0$. If the rectangular Cartesian co-ordinates of a point in Π, with respect to the axes OB and OC, are (y, z), then we can set $x = 1 - y - z$, and (by uniqueness) we have areal co-ordinates of the point. The working for Case 1 may now be carried over word for word and symbol for symbol. Thus formulae (A.8.3) and (A.8.4) still hold.
- *Case 3, ABC is an obtuse-angled triangle*
  Suppose, without loss of generality, that the obtuse angle is at A. Now $a^2 > b^2 + c^2$, so we make the following substitutions: $d^2 = \frac{1}{2}(a^2 - b^2 - c^2)$, $e^2 = \frac{1}{2}(c^2 + a^2 - b^2)$, and $f^2 = \frac{1}{2}(a^2 + b^2 - c^2)$. Then

$$\begin{aligned} e^2 + f^2 &= a^2\,, \\ d^2 + b^2 &= f^2\,, \\ d^2 + c^2 &= e^2\,. \end{aligned} \tag{A.8.5}$$

These equations mean that a suitable origin O exists, which is on the normal to Π through A such that $\angle BOC = \angle OAB = \angle OAC = 90°$. We do not have a right triad of vectors, but we do have a tetrahedron $OABC$ with three right angles, one at the vertex O and two at the vertex A. Such a construction is always possible for an obtuse-angled triangle. Writing $\boldsymbol{OA} = \boldsymbol{x}$, $\boldsymbol{OB} = \boldsymbol{y}$, and $\boldsymbol{OC} = \boldsymbol{z}$, as usual, we now have $|\boldsymbol{x}|^2 = d^2$, $|\boldsymbol{y}|^2 = e^2$, $|\boldsymbol{z}|^2 = f^2$, $\boldsymbol{y} \cdot \boldsymbol{z} = 0$, and $\boldsymbol{z} \cdot \boldsymbol{x} = \boldsymbol{x} \cdot \boldsymbol{y} = d^2$.

Now, with P and Q having areal co-ordinates as defined in Case 1, so that $\boldsymbol{PQ} = u\boldsymbol{x} + v\boldsymbol{y} + w\boldsymbol{z}$, with $u + v + w = 0$, we have

$$\begin{aligned} PQ^2 &= u^2d^2 + v^2e^2 + w^2f^2 + 2(uv + uw)d^2 \\ &= (uv + uw)d^2 + v^2e^2 + w^2f^2 \\ &= \frac{1}{2}\left[(uv + uw)(a^2 - b^2 - c^2) + v^2(c^2 + a^2 - b^2) + w^2(a^2 + b^2 - c^2)\right] \\ &= \frac{1}{2}\left[a^2(uv + uw + v^2 + w^2) + b^2(w^2 - v^2 - uv - uw)\right. \\ &\qquad\qquad\qquad\qquad\qquad\left. + c^2(v^2 - w^2 - uv - uw)\right] \\ &= -a^2vw - b^2wu - c^2uv\,, \end{aligned} \tag{A.8.6}$$

which is the same as formula (A.8.3) found in Cases 1 and 2.
An alternative form, which may be useful if arc lengths of curves are to be calculated, is very easily deduced:

$$PQ^2 = u^2bc\cos A + v^2ca\cos B + w^2ab\cos C\,. \tag{A.8.7}$$

However, the form (A.8.6) is more useful in geometrical applications.

We now give a worked example to illustrate the above.

Example A.8.1 We compute I_1I_2, the distance between the excentres of the excircles opposite A and B. The co-ordinates of these points are $I_1(-a, b, c)/(b + c - a)$ and $I_2(a, -b, c)/(c + a - b)$. Hence $\boldsymbol{I_1I_2} = [2c/(c + a - b)(b + c - a)]\,(a, -b, b - a)$. Using eqn (A.8.6) for the displacement $\boldsymbol{PQ} = (a, -b, b - a)$, we obtain

$$\begin{aligned} PQ^2 &= a^2b(b - a) - b^2a(b - a) + c^2ab \\ &= ab(c + a - b)(c + b - a)\,. \end{aligned} \tag{A.8.8}$$

Hence

$$I_1I_2 = 2c\sqrt{\frac{ab}{(c + a - b)(b + c - a)}}\,. \tag{A.8.9}$$

A.9 The condition for perpendicular displacements

Let $\boldsymbol{PQ} = (u, v, w)$ and $\boldsymbol{RS} = \boldsymbol{PT} = (f, g, h)$ be two displacements in the plane Π, so that $u + v + w = f + g + h = 0$. The condition that $\boldsymbol{PQ}$ and $\boldsymbol{RS}$ should be perpendicular is

$$PQ^2 + PT^2 = QT^2\,, \tag{A.9.1}$$

where $\boldsymbol{QT} = (f - u, g - v, h - w)$. Using eqn (A.8.6) three times, the condition (A.9.1) becomes

$$-a^2vw - b^2wu - c^2uv - a^2gh - b^2hf - c^2fg$$
$$= -a^2(g-v)(h-w) - b^2(h-w)(f-u) - c^2(f-u)(g-v),$$

and this reduces to

$$a^2(gw+hv) + b^2(hu+fw) + c^2(fv+gu) = 0\,. \qquad \text{(A.9.2)}$$

The left-hand side of eqn (A.9.2) may be thought of as twice the negative of the scalar product of the two displacements (u, v, w) and (f, g, h), and then the right-hand side of eqn (A.8.6) is the scalar product of (u, v, w) with itself.

Example A.9.1 We show how to calculate the condition that the two lines with equations $px+qy+rz=0$ and $lx+my+nz=0$ are perpendicular. A displacement along the first line has co-ordinates $(u, v, w) = (q-r, r-p, p-q)$, and a displacement along the second line has co-ordinates $(f, g, h) = (m-n, n-l, l-m)$. The vanishing of the scalar product of these two displacements (with the scalar product defined as in the previous paragraph) now gives the required result.

Example A.9.2 Let $P(l, m, n)$ be a point in the plane Π, not at the vertices of the triangle ABC and not lying on BC, so that $l \neq 0$. Let the foot of the perpendicular from P onto BC be $L(0, q, r)$. We work out the distances BL and LC. We have $\boldsymbol{BC} = (0, -1, 1)$ and $\boldsymbol{PL} = (-l, q-m, r-n)$. Using eqn (A.9.2), with $u=0$, $v=-1$, and $w=1$, and $f=-l$, $g=q-m$, and $h=r-n$, we obtain

$$a^2(q-m-r+n) + b^2(-l) + c^2l = 0\,. \qquad \text{(A.9.3)}$$

Hence $q-r = m-n+(b^2-c^2)l/a^2$. However, $q+r=1$ and $l+m+n=1$, and so

$$q = m + \frac{(a^2+b^2-c^2)l}{2a^2},$$
$$r = n + \frac{(c^2+a^2-b^2)l}{2a^2}, \qquad \text{(A.9.4)}$$

that is, $q = m + (bl/a)\cos C$ and $r = n + (cl/a)\cos B$. It follows that

$$BL = an + cl\cos B\,,$$
$$LC = am + bl\cos C\,. \qquad \text{(A.9.5)}$$

These equations are used in Section 8.3.

A.10 The equation of a circle

Now that we can measure distances we can obtain the equation of a circle. We suppose that the centre of the circle is the point $P(l, m, n)$ and that (x, y, z) is any point on the circumference. Let the radius of the circle be R. Then, using eqn (A.8.6), the equation of the circle is

$$-a^2(y-m)(z-n)-b^2(z-n)(x-l)-c^2(x-l)(y-m)=R^2\,. \qquad \text{(A.10.1)}$$

However, this is not the appropriate form of the equation, as we need to make it a homogeneous equation of degree 2 in x, y, and z.

This is done by using the condition $x+y+z=1$. Then eqn (A.10.1) becomes

$$\begin{aligned}-a^2\,[(n+l)y-m(z+x)]\,[(l+m)z-n(x+y)]\\ -b^2\,[(l+m)z-n(x+y)]\,[(m+n)x-l(y+z)]\\ -c^2\,[(m+n)x-l(y+z)]\,[(n+l)y-m(z+x)]=R^2(x+y+z)^2\,.\end{aligned} \qquad \text{(A.10.2)}$$

Equation (A.10.2) reduces to

$$ux^2+vy^2+wz^2+2pyz+2qzx+2rxy=0\,, \qquad \text{(A.10.3)}$$

where

$$u=-R^2+b^2n^2+c^2m^2+(b^2+c^2-a^2)mn\,, \qquad \text{(A.10.4)}$$

with similar expressions for v and w by cyclic changes of l, m, and n, and a, b, and c, and where

$$\begin{aligned}2p&=-2R^2-a^2l^2-a^2lm-a^2nl-2a^2mn\\ &\qquad+b^2l^2+b^2lm-b^2nl+c^2l^2-c^2lm+c^2nl \qquad \text{(A.10.5)}\\ &=-2R^2+(b^2+c^2-a^2)l^2-2a^2mn-(a^2+b^2-c^2)nl-(c^2+a^2-b^2)lm\,,\end{aligned} \qquad \text{(A.10.6)}$$

with similar expressions for $2q$ and $2r$ by cyclic changes of letters.

Now,

$$\begin{aligned}\frac{v+w-2p}{a^2}&=l^2+m^2+n^2+2mn+2nl+2lm\\ &=(l+m+n)^2=1\,,\end{aligned}$$

and hence it follows, by symmetry, that

$$\frac{v+w-2p}{a^2}=\frac{w+u-2q}{b^2}=\frac{u+v-2r}{c^2}\,. \qquad \text{(A.10.7)}$$

What we have proved is that every circle in areal co-ordinates has an equation of the form (A.10.3) subject to the side conditions (A.10.7). The converse is true, that, if an equation homogeneous of degree 2 in x, y, and z is of the form (A.10.3), subject to the side conditions (A.10.7), then it represents a circle. The proof is left to the reader. Equation (A.10.3), in general, represents a conic.

Theorem A.10.1 *The equation of the tangent at (x_0, y_0, z_0) to the conic with equation (A.10.3) is*

$$(ux_0 + ry_0 + qz_0)x + (rx_0 + vy_0 + pz_0)y + (qx_0 + py_0 + wz_0)z = 0\,. \quad (A.10.8)$$

□

This result is proved by the usual calculus methods and is left to the reader.

We now obtain the equations of the key circles associated with a triangle.

The circumcircle

The circumcircle passes through $A(1, 0, 0)$, $B(0, 1, 0)$, and $C(0, 0, 1)$. Hence $u = v = w = 0$. From eqns (A.10.7) we may put $2p = a^2$, $2q = b^2$, and $2r = c^2$, so that the equation of the circumcircle is

$$a^2yz + b^2zx + c^2xy = 0\,. \quad (A.10.9)$$

Assume now that the triangle ABC is not isosceles. The equation of the tangent at A, by eqn (A.10.8), is $c^2y + b^2z = 0$, and this meets BC at the point L with co-ordinates $L(0, b^2, -c^2)/(b^2 - c^2)$. If M and N are similarly defined in terms of the tangents at B and C, then clearly L, M, and N are collinear on the line with equation $x/a^2 + y/b^2 + z/c^2 = 0$. If the harmonic conjugates of L, M, and N are denoted by L', M', and N', respectively, then L' has co-ordinates $L'(0, b^2, c^2)/(b^2 + c^2)$, with similar expressions for M' and N'. It is now evident that AL', BM', and CN' are concurrent at the point $S(a^2, b^2, c^2)/(a^2 + b^2 + c^2)$. S is the symmedian point, which appears in Section 8.5 as the isogonal conjugate of the centroid G.

Assume now that the triangle ABC does not have a right angle. If the tangents at B and C to the circumcircle meet at L'', with M'' and N'' similarly defined, then it is easy to show that L'' has co-ordinates $(-a^2, b^2, c^2)$, with similar expressions by cyclic changes for M'' and N''. Then AL'', BM'', and CN'' are concurrent at S, which is yet another way of defining the symmedian point.

If a point P has areal co-ordinates (l, m, n) then its *isogonal conjugate* Q has unnormalised areal co-ordinates $(a^2/l, b^2/m, c^2/n)$. Obviously, the isogonal conjugate of Q is P.

The incircle

We claim that the equation of the incircle is

$$(s-a)^2x^2 + (s-b)^2y^2 + (s-c)^2z^2 - 2(s-b)(s-c)yz - 2(s-c)(s-a)zx - 2(s-a)(s-b)xy = 0\,, \quad (A.10.10)$$

where $s = \frac{1}{2}(a + b + c)$.

In fact,

$$\frac{v+w-2p}{a^2} = \frac{(s-b)^2+(s-c)^2+2(s-b)(s-c)}{a^2} = 1$$
$$= \frac{w+u-2q}{b^2} = \frac{u+v-2r}{c^2},$$

so eqn (A.10.10) represents a circle. Also, it touches the line $x = 0$ at the point with co-ordinates $(0, s-c, s-b)/a$, and similarly it touches the lines $y = 0$ and $z = 0$. The only circle with these properties is the incircle.

The excircle opposite A

We leave it as an exercise for the reader to show that the equation of the excircle opposite A is

$$s^2x^2+(s-c)^2y^2+(s-b)^2z^2-2(s-b)(s-c)yz+2s(s-b)zx+2s(s-c)xy = 0\,. \tag{A.10.11}$$

The nine-point circle

Incorporating the conditions (A.10.7) and the fact that the nine-point circle passes through the midpoints of the sides $L(0, \frac{1}{2}, \frac{1}{2})$, $M(\frac{1}{2}, 0, \frac{1}{2})$, and $N(\frac{1}{2}, \frac{1}{2}, 0)$, we find its equation to be

$$(b^2+c^2-a^2)x^2+(c^2+a^2-b^2)y^2+(a^2+b^2-c^2)z^2-2a^2yz-2b^2zx-2c^2xy = 0\,. \tag{A.10.12}$$

The polar circle

The polar circle is the circle with respect to which A is the pole of BC, B is the pole of CA, and C is the pole of AB. Its equation is

$$\cot A\, x^2 + \cot B\, y^2 + \cot C\, z^2 = 0\,. \tag{A.10.13}$$

Exercises A

A.1 Show that the equation of the circle which is tangent to AB at B and which passes through C is $c^2x^2 - a^2yz + (c^2-b^2)zx = 0$. Show that the circle which is tangent to BC at C and which passes through A meets the first circle at the Brocard point Ω, whose unnormalised areal co-ordinates are $\Omega(1/c^2, 1/a^2, 1/b^2)$. What is the isogonal conjugate of Ω?

A.2 In the triangle ABC let P have areal co-ordinates (l, m, n). Prove that

$$AP^2 = c^2m^2 + b^2n^2 + 2mnbc\cos A\,.$$

A.3 Let L, M, and N be the feet of the perpendiculars from a point P onto the sides BC, CA, and AB, respectively. Prove that AL, BM, and CN are concurrent or parallel if and only if P lies on a cubic curve passing through O, H, I, A, B, C, I_1, I_2, and I_3.

A.4 Let P have areal co-ordinates (l, m, n) and suppose that L, M, and N are the feet of the perpendiculars from P onto the sides BC, CA, and AB, respectively. Prove that the lines through A, B, and C perpendicular to the sides MN, NL, and LM, respectively, of the pedal triangle LMN are concurrent at the isogonal conjugate Q, whose unnormalised areal co-ordinates are $(a^2/l, b^2/m, c^2/n)$.

A.5 Show that the equation of the circle $I_1I_2I_3$ is

$$bcx^2 + cay^2 + abz^2 + (a + b + c)(ayz + bzx + cxy) = 0\,.$$

A.6 Prove Feuerbach's theorem (see Section 8.9) that the nine-point circle touches the inscribed and escribed circles at the points P, P_1, P_2, and P_3, and that the tangents at these points have equations

$$\frac{x}{b-c} + \frac{y}{c-a} + \frac{z}{a-b} = 0$$

together with the three equations formed from this by changing a into $-a$, b into $-b$, and c into $-c$. See also Exercise A.9.

A.7 Within a triangle ABC we take two points U and V. The lines AU, BU, and CU meet the opposite sides at L, M, and N, respectively. The points of intersection of VA and MN, VB and NL, and VC and LM are denoted by D, E, and F, respectively. Prove that LD, ME, and NF are concurrent at a point P which remains the same if U and V are interchanged.

A.8 Let AL, BM, and CN, and AD, BE, and CF be two sets of Cevians in the triangle ABC. Prove that a conic passes through L, M, N, D, E, and F.

A.9 With the notation of Section A.10, prove that the equation of any circle may be expressed in the form $a^2yz + b^2zx + c^2xy - (x+y+z)(ux+vy+wz) = 0$, where u, v, and w are the powers of A, B, and C, respectively, with respect to the circle. Hence show that the common chord of two circles, one with powers u, v, and w and the other with powers u', v', and w', has equation $(u - u')x + (v - v')y + (w - w')z = 0$.

Answers to exercises

Exercises 1.1

1.1.1 (a) When $v = 1$, we have $c = a + 2$ and $b^2 = 4(a + 1)$.

(b) When $u = v + 1$, we have $a = 2v + 1$, $b = 2v^2 + 2v$, $c = 2v^2 + 2v + 1$, $c = b + 1$, and $b + c = a^2$.

1.1.2 One of u and v must be even. One of u, v, $u + v$, and $u - v$ must be divisible by 3. One of u, v, $u + v$, $u - v$, and $u^2 + v^2$ must be divisible by 5.

1.1.3 $(87, 116, 145)$, $(105, 100, 145)$, $(143, 24, 145)$, and $(17, 144, 145)$.

1.1.4 If $c = u^2 + v^2$ then $c^2 = (u^2 - v^2)^2 + (2uv)^2$. So $u^2 - v^2$ is odd and $2uv$ is even, and they are coprime.

1.1.5 See an article in the July 1996 issue of *Mathematical Gazette* (Bradley, 1996).

1.1.6 The solutions of $|a - b| = 1$ arise from $x^2 - 2y^2 = +1$ or -1, where $x = u - v$ and $y = v$. These are $x_0 = 1$ and $y_0 = 1$, and $x_{n+1} = x_n + 2y_n$ and $y_{n+1} = x_n + y_n$, for non-negative integers n. This leads to $u_0 = 2$ and $v_0 = 1$ and the triple $(3, 4, 5)$, $u_1 = 5$ and $v_1 = 2$ and the triple $(21, 20, 29)$, and $u_2 = 12$ and $v_2 = 5$ and the triple $(119, 120, 169)$. There is clearly an infinite sequence of such triangles. Since we obtain triangles that are closer and closer to being isosceles right-angled triangles, we obtain rational approximations closer and closer to $\sqrt{2}$, such as $10/7$, $58/41$, and $338/239$, the last of these already being correct to four decimal places.

1.1.7 Yes. Put one point at $(1, 0)$; then where must the rest be?

Exercises 1.2

1.2.1 The pair (u, v) and (v, u) have the same value for c, and the values of a and b are interchanged.

1.2.2 Consider a triangle with given a, b, and c, and $C = 120°$. At the vertex C extend BC by a distance b to a point C'. Now join AC'. Since the triangle ACC' is equilateral of side b, it follows that in the triangle ABC' the angle $C' = 60°$ and, whereas b and c are unchanged, the length of BC' is $a + b$.

1.2.3 $[39, 65, 91]$, $[56, 49, 91]$, $[11, 85, 91]$, and $[80, 19, 91]$.

Exercises 1.3

1.3.1 One method is to use $\cos A = -\cos(B + C)$ and $\sin B \sin C = h^2/mn$.

1.3.2 If the sides are $N-1$, N, and $N+1$ then Heron's formula gives

$$[ABC] = \sqrt{\left(\frac{1}{4}N^2 - 1\right)\frac{3}{4}N^2}\,.$$

Clearly, N must be even. So putting $N = 2M$ we obtain

$$[ABC] = M\sqrt{3(M^2-1)}\,.$$

It follows that an integer P must exist so that $M^2 - 3P^2 = 1$. This is Pell's equation and there are an infinite number of solutions, given by $M_0 = 2$ and $P_0 = 1$, and $M_{n+1} = 2M_n + 3P_n$ and $P_{n+1} = M_n + 2P_n$. The first three such triangles have sides $(3,4,5)$, $(13,14,15)$, and $(51,52,53)$.

1.3.3 There are twenty-five Heron triangles with height 40 formed from five Pythagorean triangles. Listing the base last, these are as follows:

$$(50,50,60)\,,\quad (104,104,192)\,,\quad (85,85,150)\,,\quad (58,58,84)\,,$$
$$(41,41,18)\,,\quad (104,50,126)\,,\quad (104,50,66)\,,\quad (85,50,105)\,,$$
$$(85,50,45)\,,\quad (58,50,72)\,,\quad (58,50,12)\,,\quad (41,50,39)\,,$$
$$(41,50,21)\,,\quad (85,104,171)\,,\quad (85,104,21)\,,\quad (58,104,138)\,,$$
$$(58,104,54)\,,\quad (41,104,105)\,,\quad (41,104,87)\,,\quad (58,85,117)\,,$$
$$(58,85,33)\,,\quad (41,85,84)\,,\quad (41,85,66)\,,\quad (58,41,51)\,,\quad (58,41,33)\,.$$

1.3.4 For example, $(289, 784, 975)$ and $(841, 1369, 1122)$. A construction that gives an infinite number of such triples arises from solutions of the Diophantine equation $u^4 - v^4 = p^4 - q^4$, whose common value may be taken as the altitude of a Heron triangle, two of whose sides are $(u^2+v^2)^2$ and $(p^2+q^2)^2$, the third side being $2uv(u^2+v^2) + 2pq(p^2+q^2)$. It is known that $u = 158$, $v = 133$, $p = 134$, and $q = 59$ is the solution with the lowest values of u, v, p, and q.

1.3.5 Consider the component right-angled triangles from which the Heron triangle is composed. There are several cases.

Exercises 1.4

1.4.1 $(15, 12, 16, 25)$ and $(9, 12, 20, 25)$.

1.4.2 The other value of $\tan x = v'/u'$, where $v' = u(u^2 - 3v^2)$ and $u' = v(3u^2 - v^2)$, and it follows that $u'^2 - v'^2 = x(3y^2 - x^2)$ and $2u'v' = y(3x^2 - y^2)$. Also, $u'^2 + v'^2 = (u^2+v^2)^3$.

1.4.3 We have $a = 2x+1$, $b = 2x$, $c = 4x^2 + 2x$, and $d = 4x^2 + 2x + 1$, and $a = 2x - 1$, $b = 2x$, $c = 4x^2 - 2x$, and $d = 4x^2 - 2x + 1$, where x is any positive integer.

1.4.4 This provides an efficient, but not systematic, method of finding Pythagorean quartets. See Section 7.5 for a parametric solution of this equation.

Exercise 1.5

1.5.1 The eighteen triangles are as follows:

$$(305, 289, 18), \quad (170, 153, 19), \quad (104, 85, 21), \quad (74, 51, 25),$$
$$(65, 34, 33), \quad (97, 90, 11), \quad (58, 50, 12), \quad (40, 30, 14), \quad (37, 26, 15),$$
$$(34, 20, 18), \quad (78, 75, 9), \quad (39, 35, 10), \quad (30, 25, 11), \quad (24, 15, 15),$$
$$(41, 40, 9), \quad (26, 24, 10), \quad (20, 16, 12), \quad (15, 14, 13).$$

Exercises 1.6

1.6.1 In an equable parallelogram we have $a = p(k-1)$, $b = p$, and $h = 2k$, and $a = b$ implies that $k = 2$ and $h = 4$.

1.6.2 In an equable rectangle $h = a$, so $ab = 2(a+b)$ or $(a-2)(b-2) = 4$, and so the only possibilities are $a = b = 4$, or $a = 6$ and $b = 3$, or $a = 3$ and $b = 6$.

1.6.3 4.

1.6.4 Try $a = u + v + 1$ and $b = u - v$. What extra condition has to be imposed?

Exercises 2.1

2.1.1 The maximum value of $v(u-v)$ subject to fixed $u^2 + v^2$ is when $u = v(1+\sqrt{2})$, and this leads to $h = 1 + \sqrt{2}$, and this is the best possible since rational u and v exist so that u/v can be made arbitrarily close to $1 + \sqrt{2}$.

2.1.2 There are thirteen Heron triangles with $r = 3$. These have sides

$$(11, 100, 109), \quad (7, 65, 68), \quad (12, 55, 65), \quad (13, 40, 51),$$
$$(8, 26, 30), \quad (7, 24, 25), \quad (15, 28, 41), \quad (16, 25, 39), \quad (8, 15, 17),$$
$$(19, 20, 37), \quad (11, 13, 20), \quad (9, 12, 15), \quad (10, 10, 12).$$

2.1.3 The maximum value is $h = 2$. This is because $R = 2r$ for an equilateral triangle and integer-sided triangles exist that are as close to being equilateral as one desires. For example, it was shown in Exercise 1.3.2 that there are indefinitely large integer values of N for which $(N-1, N, N+1)$ is a Heron triangle.

2.1.4 This would require that $u^2 + v^2 = 2hv(u-v)$, where u and v are coprime and of opposite parity.

Exercise 2.2

2.2.1 There are, in fact, five possibilities. These are $t = 40$, $h = 104$, and $R = 96$. Then there are four possible chords C_1, C_2, C_3, and C_4, with $a_1 = 32$ and $b_1 = 50$, $a_2 = 20$ and $b_2 = 80$, $a_3 = 16$ and $b_3 = 100$, and $a_4 = 10$ and $b_4 = 160$. It is also possible that C_4 is a diameter and then $t = 40$, $h = 85$, and $R = 75$, so then C_1, C_2, and C_3 are possible chords. It is also possible that

C_3 is a diameter and then $t = 40$, $h = 58$, and $R = 42$, so then C_3 and C_4 are possible chords. It is also possible that C_2 is a diameter with $t = 40$, $h = 50$, and $R = 30$, so then C_1 is a possible chord. If C_1 is a diameter with $t = 40$, $h = 41$, and $R = 9$ then there is just a tangent and a diameter and no other chord.

Exercises 2.3

2.3.1 No. A quadrilateral may not exist with these values for its sides and diagonals. For a start, one must have $a + b + c > d$, $e > a + d$, etc. However, even if the eight conditions of this type are satisfied, a quadrilateral only has eight degrees of freedom; two of these are translational and one rotational, leaving **five** internal degrees of freedom only. Hence **six** lengths cannot be chosen arbitrarily to form the sides and diagonals of a quadrilateral.

2.3.2 This comes from the cosine rules for the triangles ABD and BCD and the fact that $\cos C = -\cos A$.

2.3.3 Using the formulae in the text, we have $r^2 = 19/4$.

2.3.4 Yes. We have

$$\begin{aligned} A_1A_2 &= 1\,, \\ A_2A_3 &= 3\,, \\ A_3A_4 &= 5\,, \\ &\vdots \\ A_{1000}A_{1001} &= 1999\,, \\ A_{1001}A_{1002} &= 2000\,, \\ &\vdots \\ A_{2000}A_1 &= 2\,. \end{aligned}$$

2.3.5 $AB = 18$ and $BC = 10$. BQ is also 10.

Exercise 2.4

2.4.1 $a = c = 2k|2q^2 - p^2|$, $b = 4kpq$, and $n = l = k(p^2 + 2q^2)$.

Exercise 2.5

2.5.1 (i) $r = 4$, $r_1 = 5$, $r_2 = 24$, and $r_3 = 120$.

(ii) $r = 4$, $r_1 = 8$, $r_2 = 9$, and $r_3 = 72$.

Exercise 2.6

2.6.1 $q = 502 = 6k+4$, where $k = 83$. So the number of triangles is $(502^2 - 4)/12 = 21\,000$.

Exercise 2.7

2.7.1 $k = 5$ and $CW = 80$.

Exercises 2.8

2.8.1 Yes, the external bisectors are rational. The formula for the length of the external bisector of angle A is $AU'^2 = bc(a-b+c)(a+b-c)/|b-c|^2$, and in the triangle with sides 125, 169, and 154 the lengths are $AU' = 975/2$, $BV' = 23\,100/29$, and $CW' = 4004/3$.

2.8.2 The sines of the half-angles are not rational. The component right-angled triangles can only be fitted together in one way to give rational sines. The wrong way round, you get an annoying $\sqrt{2}$ appearing in the analysis.

Exercises 2.9

2.9.1 $p = 5$ and $q = 3$ gives $a = 27$, $b = 48$, and $c = 35$. Another possibility is $p = 3$ and $q = 2$ when enlarged by a factor of 6. This gives $a = 48$, $b = 60$, and $c = 18$.

2.9.2 For $n = 5$ we have

$$a = q^5\,, \quad b = q(p^4 - 3p^2q^2 + q^4)\,, \quad c = p(p^2 - q^2)(p^2 - 3q^2)\,.$$

Then $p = 7$ and $q = 4$ gives $a = 1024$, $b = 1220$, and $c = 231$. For $n = 6$ we have

$$a = q^6\,, \quad b = pq(p^2 - q^2)(p^2 - 3q^2)\,, \quad c = p^2(p^2 - 2q^2)(p^2 - 3q^2) - q^6\,.$$

Then $p = 11$ and $q = 6$ gives $a = 46\,656$, $b = 72\,930$, and $c = 30\,421$.

Exercises 3.1

3.1.1 $[OAB] = 12$, $B = 12$, and $I = 7$.

3.1.2 $[OAB] = 12$, $B = 8$, and $I = 9$.

3.1.3 $[OAB] = 3$, $B = 6$, and $I = 1$.

3.1.4 $[OAB] = 10$, $B = 6$, and $I = 8$.

3.1.5 $[OAB] = 2$, $B = 6$, and $I = 0$.

3.1.6 The number of internal lattice points is approximately equal to the area, so this number is 62 800 to three significant figures.

Exercise 3.2

3.2.1 $I = 39$ and $B = 11$, so the area is 43.5.

Exercises 3.3

3.3.1 $x = -1 + 12k$ and $y = 3 - 12k$.

3.3.2 7 divides 133 and 84, but 7 does not divide 1.

3.3.3 h must be a multiple of 7. That $h = 7$ follows from the solution $x = -5$ and $y = 8$.

3.3.4 Solutions exist when $N = 0, 4, 7, 8, 11, 12, 14, 15$, and 16, and for all $N > 17$.

Exercises 4.1

4.1.1 $x = 8 - 2(m + 8)/(m^2 + 1)$ and $y = -1 + 2(1 - 8m)/(m^2 + 1)$, where m is any rational. $m = 0$ gives $(-8, 1)$, m tends to infinity gives $(8, -1)$, $m = 1$ gives $(-1, -8)$, $m = -1$ gives $(1, 8)$, $m = 2$ gives $(4, -7)$, $m = 5$ gives $(7, -4)$, $m = -3$ gives $(7, 4)$, $m = -8$ gives $(8, 1)$ (this line is tangent at $(8, 1)$), $m = 1/3$ gives $(-7, -4)$, and $m = -1/5$ gives $(-7, 4)$.

4.1.2 The rational points are given by

$$(x, y) = \left(\frac{7u^2 + v^2}{7u^2 - v^2}, \frac{2uv}{7u^2 - v^2}\right)$$

and the integer points are given by $(x_m, y_m) = (8, 3)A^m$, where A is the matrix with first row $(8, 3)$ and second row $(21, 8)$, and $m = 0, 1, 2, \ldots$, together with $(1, 0)$ and all of their negative counterparts. There are no integer solutions to the equation $7x^2 - y^2 = 1$ since the left-hand side is 0, 2, or 3 (mod 4) and cannot therefore equal 1.

4.1.3 $x = 17$ and $y = 4$. No, there is no solution of $x^2 - 18y^2 = -1$ because it would have to have $y = 1$, 2, or 3, and it is easily checked that none of these work.

4.1.4 $(7, 3)$, $(7, -3)$, $(-7, 3)$, and $(-7, -3)$.

4.1.5 The conic degenerates into the two straight lines $y = 2x + 3$ and $y + 2x + 3 = 0$.

Exercises 4.2

4.2.1 Any integer points on the line $x + y = 0$, and $(1, 2)$, $(2, 1)$, $(1, 0)$, $(0, 1)$, and $(2, 2)$.

4.2.2 The equations for the singular point produce a quadratic in x with integer coefficients. Since real roots occur in conjugate pairs, it is not possible for an irrational root to be a double point. It follows that the double point is rational, and $p(x) = 0$ has this double root.

4.2.3 $x = m^2 - 3$ and $y = m(m^2 - 3)$, where m is an integer, and $x = 0$ and $y = 0$.

Exercises 4.3

4.3.1 The line joining $E(0, -3)$ and $D(-1, -2)$ has equation $y + x + 3 = 0$, and this meets the curve again at $F(2, -5)$. The line joining $B(0, 3)$ to $C(2, 5)$ has equation $y = x + 3$, and this meets the curve again at $A(-1, 2)$.

4.3.2 The equation of the line joining $B_0(801/4, -22\,671/8)$ and $D(-1, -2)$ is given by $14y + 197x + 225 = 0$, and this meets the curve again at the point $(-61/49, 496/343)$.

4.3.3 We have

$$P_0 + P_0 = E * (P_0 * P_0) = E * P_1 = Q_1\,.$$

Now

$$Q_1 + P_1 = E * (P_1 * Q_1) = E * E = E\,,$$

and hence $Q_1 = -P_1$. Similarly, $Q_k = -P_k$. Now

$$3P_0 = P_0 + 2P_0 = E * (P_0 * 2P_0) = (-17 : 73 : 38)\,.$$

More generally, by induction, we can find $nP_0 = E * (P_0 * P_{n-1})$.

4.3.4 The common co-ordinates are $(-2, -1)$.

4.3.5 If C is collinear with A and B, then $C = A * B$ and

$$\begin{aligned} A + B + A * B &= A + E * (B * (A * B)) \\ &= A + E * A \\ &= E * (A * (E * A)) \\ &= E * E\,. \end{aligned}$$

Uniqueness ensures the converse.

4.3.6 If $A = B * B$ then

$$\begin{aligned} A + 2B &= B * B + E * (B * B) \\ &= E * ((B * B) * (E * (B * B))) \\ &= E * E\,. \end{aligned}$$

4.3.7 In the projective plane, $X^3 + Y^3 = Z^3$ contains the integer points $A(1 : 0 : 1)$, $B(0 : 1 : 1)$, and $J(1, -1, 0)$, and no other integer points, by Fermat's last theorem for $n = 3$. We have $A * B = J$, $A * A = A$, $B * B = B$, and $J * J = J$. With J as the origin, we have $A + A = B$, $A + B = J$, and $B + B = A$, so the group of rational points is finite, the cyclic group of order 3.

4.3.8 For all k other than $k = 0$, 8, or -8.

Exercises 4.4

4.4.1 $(0, 0)$.

4.4.2 $(0,0)$, $(2,0)$, and $(-2,0)$.

4.4.3 As in Example 4.4.2 of the text, $x = 1 \pmod 4$ and y is even. Now $y^2 + 4 = (x+3)(x^2 - 3x + 9)$ and the quadratic is 3 (mod 4), and must therefore have a prime factor that is 3 (mod 4). However, $y^2 + 4$ can have no such factor. Here we are using the result that, if $a^2 + b^2$ has a factor $k = 3 \pmod 4$, then $k|a$ and $k|b$.

4.4.4 From the text, the condition is $2s|(3r^2 + a)$.

Exercises 5.1

5.1.1 If p is an odd prime of the form $4k + 1$ then there exists an integer n such that $4T_n + 1 = 0 \pmod p$.

5.1.2 $2x^2y^2 = 8(T_l * T_m) + 1$ and $x^4 = 8T_{2l(l+1)} + 1$.

5.1.3 $(6^2 + 3^2)(6^2 + 1^2) = 39^2 + 12^2$.

5.1.4 $l = m = 2k + 1$ will serve.

5.1.5 $l = 11$, then $4T_l + 1 = 265 = 5 \times 53$.

5.1.6 $L = 63$, $M = 48$, and $2k = 56$; $2016 + 1176 = 2 \times 1596$.

Exercises 5.2

5.2.1 $x = 5$, $y = 6$, $z = 4$, and $w = 2$, and

$$x^2 + y^2 + z^2 + w^2 = 2(T_m + T_n + T_p + T_q) + 1 = 81\,.$$

5.2.2 We have

$$\begin{aligned} 87 &= 9^2 + 2^2 + 1^2 + 1^2 \\ &= 7^2 + 6^2 + 1^2 + 1^2 \\ &= 7^2 + 5^2 + 3^2 + 2^2 \\ &= 6^2 + 5^2 + 5^2 + 1^2\,. \end{aligned}$$

So

$$\begin{aligned} 43 &= T_{-3} + T_4 + T_5 + T_5 \\ &= T_1 + T_6 + T_6 \\ &= T_1 + T_3 + T_5 + T_6 \\ &= T_2 + T_3 + T_3 + T_7\,. \end{aligned}$$

5.2.3 $m = 7$, $n = -6$, $p = -1$, and $q = -1$.

Exercises 5.3

5.3.1 We require $(6n-1)^2 - 24x^2 = 1$. The first $n > 1$ satisfying this equation may be found by the method explained in Example 4.2.1 and is $n = 81$. Then $x = 99$.

5.3.2 The hexagonal numbers are

$$H_n = \frac{1}{2}n(4n-2) = n(2n-1) = \frac{1}{2}(2n-1)(2n) = T_{2n-1}\,.$$

5.3.3 The equation to be satisfied is $(72n-35+12x)(72n-35-12x) = 1225$, and this has two solutions $n = 1$ and $x = 1$, and $n = 9$ and $x = 51$.

5.3.4 By definition, $(N+1)_n - N_n = \frac{1}{2}n(n-1) = T_{n-1}$.

5.3.5 $N_{18} = 33^2$.

5.3.6 The octagonal numbers are of the form $n(3n-2)$. Now n and $3n-2$ are either coprime or they have highest common factor 2. If coprime, then both n and $3n-2$ are perfect squares. If they have highest common factor 2, and n is not a perfect square, then n is of the form $2b^2$ for some integer b. This would require $3b^2-1$ to be a perfect square also, which is not possible.

5.3.7 The next smallest is the 1456th Hex number equal to 6 355 441.

5.3.8 The next smallest is the 120th centred square number equal to 28 561.

Exercise 5.4

5.4.1 There are an infinite number of such cases. The first two non-trivial cases are $8^3 - 7^3 = 13^2$ and $105^3 - 104^3 = 181^2$.

Exercise 5.5

5.5.1 The method is to split the figure into two parts on either side of the triangle whose vertices are the labels 1, 2, m ($3 \leqslant m \leqslant n+2$). You then get the same recurrence relation as the one in the text for the problem of the $2n$-gon.

Exercises 6.1

6.1.1 We have $p = q = 37$, $r = 33$, and $s = 7$. Then $a = 1281$, $b = 319$, $c = 1138$, and $d = 1480$. Also, $\cos\theta = 969/1769$, $\sin\theta = 1480/1769$, and $[ABCD] = 341\,880$.

6.1.2 With $u = 140$, $v = 1221$, $h = 660$, and $k = 259$, then doubling p and s for parity considerations we obtain $p = 3\,020\,882$, $q = 502\,681$, $r = 34\,188$, and $s = 683\,760$. Then

$$\begin{aligned} a &= 1\,220\,684\,915\,521\,, \\ b &= 531\,618\,937\,921\,, \\ c &= 1\,284\,776\,115\,842\,, \\ d &= 1\,376\,492\,298\,720\,, \end{aligned}$$

and

$$[ABCD] = 646\,063\,949\,279\,568\,100\,622\,880\,.$$

Exercises 6.2

6.2.1 $a = 35$, $b = 5$, $c = 25$, $d = 25$, and $[ABCD] = 400$.

6.2.2 $a = 720$, $b = 312$, $c = 960$, $R = 481$, and $[ABCD] = 241\,920$.

Exercise 6.3

6.3.1 $l = 7/16$, $m = 0$, and $n = 9/16$. This means that P lies on CA and coincides with M.

Exercises 6.4

6.4.1 $l = 15$, $m = 161$, and $n = 9$. Geometrically, this means that there is a point P internal to an equilateral triangle that is at distance $7/8$ from B and at distance $169/176$ from C. (This is, of course, obvious as such a point can be constructed using a measuring rod and compass.)

6.4.2 $a^2 = 2700$.

Exercise 7.1

7.1.1 The radius of C_3 is $1/9$, and the radius of C_4 is $1/25$. The radius of C_n is $1/F_n^2$, where F_n is the nth Fibonacci number in the sequence $1, 2, 3, 5, 8, \ldots$.

Exercises 7.2

7.2.1 $\rho = 33$ or 1. This means that, if we start with three circles of radii 528, 132, and 44 touching externally, then the circle in the small space between them will have radius 16. The smaller value of ρ indicates that the second circle also has a radius of 528 and touches the other three externally, with the circle of radius 44 in the space between those of radii 528 and 132. In the latter situation one can reduce the radii by a factor of 44 to give two circles of radius 12, one of radius 3, and the one between of radius 1.

7.2.2 $\alpha = 1$, $\beta = 12$, $\gamma = 88$, and $\rho = 169$ or 33. The second value of ρ means that three touching circles of radii 264, 22, and 8 have a small space between them into which a fourth circle of radius 3 can fit and touch them all.

Exercises 7.3

7.3.1 $[0,1,1,3] = 3$, $[0,1,1,4] = 0$, $[0,1,3,4] = 6$ (repeated), and $[1,1,3,4] = 9$. Also, $4 = \frac{1}{2}(0+1+1+3+[0,1,1,3])$. The next three terms are 9, 16, and 31.

7.3.2 $[1,1,1,3] = 6$, $[1,1,1,6] = 3$, $[1,1,3,6] = 9$, and $6 = \frac{1}{2}(1+1+1+3+[1,1,1,3])$. The next five terms are 10, 19, 37, 69, and 129.

7.3.3 Spheres of radii 4275, 1900, 900, 684, and 300 may be arranged to form a set of five mutually touching spheres.

7.3.4 If the spheres touch externally then its radius is 41. If the fifth sphere surrounds the other four then its radius is 725.

7.3.5 Use the formulae in the text involving m.

Exercise 7.4

7.4.1 The next term is $340/3$, so that 36, 39, 135, 135, 183, and 340 form a touching set.

Exercises 7.5

7.5.1 $u = 16$, $v = 57$, $w = 49$, $x = 67$, $l = 1/16$, $m = 1/57$, and $n = 1/49$. Multiplying up by 44 688, we obtain $l = 2793$, $m = 784$, and $n = 912$. This gives $a = m+n = 1696$, $b = n+l = 3705$, $c = l+m = 3577$, and $[ABC] = 2\,994\,096$. The height, with a as base, is $187\,131/53$. So, multiplying up by 53, we obtain $a = 89\,888$, $b = 196\,365$, and $c = 189\,581$; then $h = 187\,131$. The two right-angled triangles have sides 30 380, 187 131, and 189 581, and 59 508, 187 131, and 196 365.

7.5.2 $u = 1638$, $v = 702$, and $w = 351$. Removing the common factor 117, we obtain $u = 14$, $v = 6$, $w = 3$, $l = 1/14$, $m = 1/6$, and $n = 1/3$. Multiplying up by 42, we obtain $l = 3, m = 7, n = 14, a = 21, b = 17, c = 10$, and $[ABC] = 84$. With a as base, the height is 8.

Exercises 8.1

8.1.1 $BD = 16/3$, $DC = 10/3$, $CE = 5/3$, $EA = 4/3$, $AF = 4/3$, $FB = 8/3$, $FD = 4$, $FE = 2/3$, and $ED = 10/3$.

8.1.2 $BD = 4$, $DC = 2$, $CE = 1$, $EA = 2$, $AF = 2$, $FB = 2$, $FD = 3$, $FE = 1$, and $ED = 2$.

8.1.3 The simplified equations are

$$\begin{aligned} l &= (u+x)(u+v)\,, \\ m &= (v+x)(u+v)\,, \\ n &= x(u+v) - uv\,, \\ y &= a(u^2+uv+v^2)\,. \end{aligned}$$

Exercises 8.2

8.2.1 $u = 70$, $v = 55$, and $w = 90$. $XY = 990/13$, $YZ = 924/13$, and $ZX = 66$.

8.2.2 A two-parameter solution is $a = 4pqv$, $b = |(p^2 - q^2)u - (p^2 + q^2)v|$, and $y = |2(p^2 + q^2)uv - 2v^2(p^2 - q^2)|$. So, given rational u and v, we can find an infinite number of a and b such that all of LM, MN, and NL are rational.

Exercise 8.3

8.3.1 $l = 9/35$, $m = 13/35$, $x = 111/35$, $NL = LM = 444/175$, and $MN = 2496/875$.

Exercises 8.4

8.4.1 The pivot point is (a) the circumcentre, and (b) the orthocentre.

8.4.2 The ratios are

$$
\begin{aligned}
&[BCP] : [CAP] : [ABP] = \\
&\quad (a+b)(a+c)a^2 \left[(b+c)(b^2+c^2) + a(b^2+bc+c^2) - a^2(b+c) - a^3\right] : \\
&\quad (b+c)(b+a)b^2 \left[(c+a)(c^2+a^2) + b(c^2+ca+a^2) - b^2(c+a) - b^3\right] : \\
&\quad (c+a)(c+b)c^2 \left[(a+b)(a^2+b^2) + c(a^2+ab+b^2) - c^2(a+b) - c^3\right].
\end{aligned}
$$

8.4.3 Another proof of the six-circle theorem is given in Hahn (1994). It uses cross-ratios in the complex plane.

Exercises 8.5

8.5.1 $1\,168\,410/7397$, $1\,549\,992/12\,193$, and $5080/41$.

8.5.2 $260/21$.

Exercises 8.6

8.6.1 *Hint*: Write down vectors parallel to the internal bisectors of the angles B and C.

8.6.2 $GH = 2GO$, $IG = 2JG$, and $\angle IGH = \angle JGO$ (opposite).

8.6.3 Use the result of Exercise 8.6.1.

8.6.4 We obtain

$$
\frac{1}{a+b+c}\left(-a^3 - b^3 - c^3 + 2a^2b + 2b^2a + 2b^2c \right.
$$
$$
\left. + 2c^2b + 2c^2a + 2a^2c - 9abc\right).
$$

Exercise 8.7

8.7.1 $r_1 = 132$, $r_2 = 168$, $r_3 = 280/3$, $AI_1 = 260$, $BI_2 = 280$, $CI_3 = 728/3$, $I_2I_3 = 910/3$, $I_3I_1 = 845/3$, and $I_1I_2 = 325$.

Exercises 8.8

8.8.1 $OI = 13/8$.

8.8.2 $OH = 23/8$.

Exercises 8.9

8.9.1 $17/4$, $37/4$, $53/4$, and $193/4$.

8.9.2 The triangles IGT and WGO are similar.

Exercises 9.1

9.1.1 $a = 11$. The tetrahedron with sides 110, 99, 79, 77, 57, and 46 has rational volume.

9.1.2 48.

Exercises 9.2

9.2.1 $a = 168$, $b = 178$, $c = 158$, and $R = 103$.

9.2.2 $d = 38$, $a = 48$, $x = 26$, and $R = 361/13$.

Exercises 9.3

9.3.1
(i) Truncated tetrahedron: 4 hexagons, 4 triangles, 12 vertices, and 18 edges.
(ii) Truncated cube: 6 octagons, 8 triangles, 24 vertices, and 36 edges.
(iii) Cuboctahedron: 6 squares, 8 triangles, 12 vertices, and 24 edges.
(iv) Truncated octahedron: 8 hexagons, 6 squares, 24 vertices, and 36 edges.
(v) Small rhombicuboctahedron: 16 squares, 8 triangles, 22 vertices, and 44 edges.
(vi) Great rhombicuboctahedron: 6 octagons, 12 squares, 40 vertices, and 60 edges.
(vii) Snub cube: 6 squares, 32 triangles, 24 vertices, and 60 edges.
(viii) Icosidodecahedron: 12 pentagons, 20 triangles, 30 vertices, and 60 edges.
(ix) Truncated dodecahedron: 12 decagons, 20 triangles, 60 vertices, and 90 edges.
(x) Truncated icosahedron: 12 pentagons, 20 hexagons, 60 vertices, and 90 edges.
(xi) Small rhombicosidodecahedron: 12 pentagons, 30 squares, 20 triangles, 60 vertices, and 120 edges.
(xii) Great rhombicosidodecahedron: 12 decagons, 30 squares, 20 hexagons, 120 vertices, and 180 edges.

(xiii) Snub dodecahedron: 12 pentagons, 80 triangles, 60 vertices, and 150 edges.

9.3.2 The five-dimensional simplex has 6 vertices, 15 edges, 20 faces, 15 tetrahedra, and 6 four-dimensional simplexes. The five-dimensional cubic polytope has 32 vertices, 80 edges, 80 faces, 40 cubes, and 10 hypercubes. Its dual has 10 vertices 40 edges, 80 faces, 80 three-dimensional cells, and 32 four-dimensional cells.

Exercises 10.1

10.1.1 Let the fourth line meet the sides of the triangle ABC at L, M, and N. The pivot theorem establishes that the circles AMN, BNL, and CLM pass through a point P. Now take CLM as a second triangle for the pivot theorem, with N on LM, B on CL, and A on CM.

10.1.2 With six lines in general position there are six Clifford circles, and these circles have a common point, the Clifford point of six lines. Then with seven lines there are seven Clifford points of six lines, and they are concyclic, and so on.

Exercise 10.2

10.2.1 Σ_Q has equation

$$x^2vw + y^2wu + z^2uv - yzu(v+w) - zxv(w+u) - xyw(u+v) = 0\,.$$

Exercise 10.3

10.3.1 S_Q has equation $27yz + 32zx + 35xy = 0$. Any integer values of x and y lead to a rational value of z, given by $z = -35xy/(32x + 27y)$. Σ_Q has equation

$$20x^2 + 15y^2 + 12z^2 - 27yz - 32zx - 35xy = 0\,.$$

A two-parameter solution is

$$\begin{aligned} x &= 12q(p+q) - 15q(p-q) = 27q^2 - 3pq\,, \\ y &= 12p(p+q) - 20p(q-p) = 32p^2 - 8pq\,, \\ z &= 12(p+q)^2 + 20p(p+q) + 15q(p+q) = 32p^2 + 59pq + 27q^2\,. \end{aligned}$$

Exercises 11.1

11.1.1 Verify this using the expressions from Theorem 11.1.1.

11.1.2 From $\mathbb{P}^2(p)$ one line is removed, leaving $p^2 + p$ lines. Then $p + 1$ points are removed, leaving p^2 points. Through each point on the deleted line there pass $p+1$ lines, including the deleted line. This means that there are p parallel lines in $A_2(p)$ (meeting at each point on the deleted line). Hence there are $p + 1$ sets of parallel lines, each set containing p lines.

11.1.3 In $\mathbb{P}^2(3)$ there are thirteen points and thirteen lines. The co-ordinates of the points may be taken as follows: $A(1,0,0)$, $B(0,1,0)$, $C(0,0,1)$, $D(0,1,1)$, $E(1,0,1)$, $F(1,1,0)$, $G(0,1,2)$, $H(2,0,1)$, $I(1,2,0)$, $J(2,1,1)$, $K(1,2,1)$, $L(1,1,2)$, and $M(1,1,1)$. The lines, together with their equations, and the points lying on them are as follows: a, $x=0$, and $BCDG$; b, $y=0$, and $ACEH$; c, $z=0$, and $ABFI$; d, $y+z=0$, and $AGKL$; e, $z+x=0$, and $BHJL$; $f, x+y=0$, and $CIJK$; $g, y+2z=0$, and $ADJM$; h, $2x+z=0$, and $BEKM$; i, $x+2y=0$, and $CFLM$; j, $2x+y+z=0$, and $EFGJ$; k, $x+2y+z=0$, and $DFHK$; l, $x+y+2z=0$, and $DEIL$; and m, $x+y+z=0$, and $GHIM$. The labelling has the pleasing dual property that, for example, a contains B, C, D, and G, and A lies on b, c, d, and g. When the line c, with equation $z=0$, is deleted we obtain the affine plane $A_2(3)$. It consists of all lines except c, making twelve lines in all. The points A, B, F, and I now lie on the line at infinity. The nine remaining points may have their z co-ordinates normalised equal to 1, and then we need only tabulate their x and y co-ordinates, which are $C(0,0)$, $D(0,1)$, $E(1,0)$, $G(0,2)$, $H(2,0)$, $J(2,1)$, $K(1,2)$, $L(2,2)$, and $M(1,1)$. The lines b, d, and g are now parallel, as are the sets a, e, and h; i, j, and k; and f, l, and m. The equations of the lines are as follows: a, $x=0$; b, $y=0$; d, $y=2$; e, $x=2$; f, $x+y=0$; g, $y=1$; h, $x=1$; i, $y=x$; j, $y=x+2$; k, $y=x+1$; l, $x+y=1$; and m, $x+y=2$.

11.1.4 (a) The difference-sets are 0, 1, and 3 (mod 7), 0, 1, 3, and 9 (mod 13), and 0, 1, 6, 8, and 18 (mod 21).

(b) It is sufficient to show that two points are joined by a unique line. Suppose that the points are P_j and P_k; then we have to show that there is a unique line p_l joining them. That is, we must have a unique solution in l, x_t, and x_s of the equations $l+j=x_t$ and $l+k=x_s$ (mod n^2+n+1) for fixed j and k, i.e., of the equations $j-k=x_t-x_s$ and $l=x_t-j$ (mod n^2+n+1). Since $j-k$ is fixed, the values of x_t and x_s are uniquely determined by the first of these equations, as this is the property of a difference-set. Once x_t is known, the value of l follows uniquely from the second equation. The structure of $\mathbb{P}^2(4)$ now follows from the fact that the line p_l ($l=0,\dots,20$) contains the points P_{21-l}, P_{22-l}, P_{27-l}, P_{29-l}, and P_{39-l}, where the subscripts are mod 21, so that they lie between 0 and 20 inclusive. So, for example, p_{11} contains P_{10}, P_{11}, P_{16}, P_{18}, and P_7.
The field for $\mathbb{P}^2(4)$ is a finite complex field consisting of 0, 1, ω, and ω^2, where $\omega^3=1$ and $1+\omega+\omega^2=0$, and the working is mod 2, so that $1+\omega=\omega^2$, $\omega+\omega^2=1$, and $\omega^2+1=\omega$. The twenty-one points have co-ordinates $A(1,0,0)$, $B(0,1,0)$, $C(0,0,1)$, $D(1,1,0)$, $E(1,0,1)$, $F(0,1,1)$, $G(1,1,1)$, $H(1,\omega,0)$, $I(0,1,\omega)$, $J(\omega,0,1)$, $K(\omega,1,0)$, $L(0,\omega,1)$, $M(1,0,\omega)$, $N(\omega,\omega,1)$, $P(\omega^2,\omega^2,1)$, $Q(1,\omega,1)$, $R(\omega,1,1)$, $S(\omega^2,1,1)$, $T(1,\omega^2,1)$, $U(\omega^2,\omega,1)$, and $V(\omega,\omega^2,1)$. The lines are $BCFIL$, $ACEJM$, $ABDHK$, $CDGNP$, $BEGQT$, $AFGRS$,

$DEFUV$, $CKRTU$, $ALNQU$, $BMPSU$, $CHQSV$, $AIPTV$, $BJNRV$, $DIQRM$, $DJLST$, $EIKNS$, $FHMNT$, $FJKPQ$, $EHLPR$, $GHIJU$, and $GKLMV$.

References

Bradley, C. J. (1988). Triangular numbers and sums of squares. *Mathematical Gazette*, **72**, 297.

Bradley, C. J. (1996). On solutions of $m^2 + n^2 = 1 + l^2$. *Mathematical Gazette*, **80**, 404.

Bradley, C. J. (1998). Equal sums of squares. *Mathematical Gazette*, **82**, 80.

Bradley, C. J. (2002). More on Simson conics and lines. *Mathematical Gazette*, **86**, 303.

Bradley, C. J. (2005). *Euclidean geometry: from theory to problem solving*, Chapters 3–7. United Kingdom Mathematics Trust, Leeds University. In press.

Bradley, C. J. and Bradley, J. T. (1996). Countless Simson line configurations. *Mathematical Gazette*, **80**, 314.

Conway, J. H. and Guy, R. K. (1996). *The book of numbers*. Springer-Verlag, New York.

Coxeter, H. S. M. (1989). *Introduction to geometry*. Wiley, New York.

Descartes, R. (1901). The correspondence of Descartes with the Princess Elizabeth in Adam and Tannery. Oevres de Descartes, Volume IV. Paris.

Dickson, L. E. (1971). *History of the theory of numbers*, Volume II. Chelsea, New York.

Durell, C. V. (1946). *Modern geometry*. Macmillan, London.

Gardiner, A. D. (1987). *Discovering mathematics*. Clarendon Press, Oxford.

Hahn, L.-S. (1994). *Complex numbers and geometry*. The Mathematical Association of America, Washington, DC.

Jones, G. A. and Jones, J. M. (1998). *Elementary number theory*. Springer-Verlag, London.

Larson, L. C. (1983). *Problem-solving through problems*. Springer-Verlag, New York.

Niven, I., Zuckerman, H. S., and Montgomery, H. L. (1991). *An introduction to the theory of numbers*. Wiley, New York.

Pedoe, D. (1970). *Geometry: a comprehensive course*. Dover, New York.

Rose, H. E. (1988). *A course in number theory*. Clarendon Press, Oxford.

Salmon, G. (1912). *A treatise on the analytic geometry of three dimensions*. Longmans, Green and Co., London.

Sastry, K. R. S. (2003). Brahmagupta quadrilaterals: a description. *Crux Mathematicorum*, **29**, 42.

Shklarsky, D. O., Chentzov, N. N., and Yaglom, I. M. (1993). *The USSR Olympiad problem book*. Dover, New York.

Silverman, J. H. (1997). *A friendly introduction to number theory*. Prentice Hall, New Jersey.

Silvester, J. R. (2001). *Geometry ancient and modern*. Oxford University Press.

Index